T0313683

The Governance of International Migration

The Governance of International Migration

Irregular Migrants' Access to Right to Stay in Turkey and Morocco

Ayşen Üstübici

Amsterdam University Press

Cover illustration: Photo by Ayşen Üstübici

Cover design: Coördesign, Leiden
Lay-out: Crius Group, Hulshout

Amsterdam University Press English-language titles are distributed in the US and Canada by
the University of Chicago Press.

ISBN   978 94 6298 276 5
e-ISBN   978 90 4853 280 3 (pdf)
DOI   10.5117/9789462982765
NUR   747

Printed and bound by CPI Group (UK) Ltd, Croydon, CR0 4YY

To my parents
Nafiye and Ahmet

Table of contents

List of figures and tables

Acknowledgements

The genesis of this book was my PhD dissertation, which I defended at Koç University and the University of Amsterdam in 2015, it further evolved during my post-doctoral fellowship at MiReKoc and assistant professorship at Koç University. I would like to thank all of the academic and administrative staff at both institutions for providing me with intellectual homes during the fieldwork and writing stages. I would like to thank Koç University, the Bucerius PhD Scholarship Program *Settling into Motion*, the Center for Gender Studies at Koç University (KOÇKAM) and the Scientific and Technological Research Council of Turkey (TÜBİTAK) for their financial support at different stages of my doctoral research.

This book is indeed a product of the long physical, intellectual and mental journey that I have taken alongside several wonderful people to whom thanks are due. Unfortunately, I can only mention a few of them in this short piece. I would first like to thank my supervisors, Prof Ahmet İçduygu and Prof Jan Rath for their invaluable guidance. Additionally, Dr Sebastien Chauvin, Dr Özlem Altan, Prof Deniz Yükseker and Prof Mine Eder were always ready to read earlier drafts of my chapters and discuss my ideas. Conducting fieldwork in two different countries would not have been possible without the valuable help of precious people. In Morocco: Apostolos, Babacar, Fatima, Moussa, Cewad and Najat, among others, helped me immensely in navigating my way in a land where I considered myself an outsider. Without the help of Deniz Sert, Deniz Karcı, Biriz, Uğur and Fattah, and many others, fieldwork in Turkey would have been much more challenging. I am indebted to my colleagues: those in the *Settling into Motion* programme, in Morocco, at the International Migration Institute Oxford, at Koç University and at UvA with whom I had the chance to discuss different bits of my research in various formal and informal settings. I would also like to thank Eda Kiriscioglu, Judy Woods, Lara Savenije and Emrah Celik for their able assistance with editing and referencing during the transition from dissertation to book.

Above all, it has been such a relief to have a home to return to with precious old friends and beloved family at the end of every venture. My parents, Nafiye and Ahmet, my brother, Alican, my aunts Kadriye and Güzin were always there for me with their unconditional love and support. Işık, my dearest, has been supportive, comforting and engaged throughout every bit of this physical and intellectual journey. Last but not least, my deepest gratitude is to the participants of my study who gave me their valuable time and trusted me with their professional and personal experiences. Without their contribution, this book would be too dry or would not exist at all. Along with my parents, I dedicate this book to all migrants for whom the journey and the home are mostly intertwined as they seek better opportunities in life.

Finally, I would like to acknowledge that some parts of the book have appeared in different publications.
- Part of Chapter 3 was published in a shorter and substantially different form in *Geopolitics* (Üstübici 2016)
- Part of Chapter 2 in a substantially different form was published in *Migration and Development* (Üstübici 2015)
- Part of Chapter 2 was published in a different form and in Turkish in *Toplum Bilim* (Üstübici 2017).

Ayşen Üstübici
February 2018

Abbreviations

ABCDS	Association Beni Znassen for Culture Development and Solidarity
AFVIC	Association for Victims of Clandestine Migration and their Families
ALECMA	Association Lumiere sur L'Emigration Clandestine au Maghreb
AMDH	The Moroccan Association for Human Rights
ANAPEC	The Moroccan National Recruitment and Employment Agency
ARMID	Association Mediterranean Encounter for Immigration and Development
ASAM	Association for Solidarity with Asylum Seekers and Migrants
ASEM	Association for Solidarity and Mutual Aid with Migrants
ATMF	Association of Workers from Maghreb in France
CCME	The Council of the Moroccan Community Living Abroad
CMSM	Council of Sub-Saharan Migrants in Morocco
CNDH	National Council of Human Rights
CSOs	Civil society organizations
DGMM	Directorate General of Migration Management
DRC	Democratic Republic of Congo
ECHR	European Convention on Human Rights
ECtHR	European Court of Human Rights
EU	European Union
FOO	Foundation Orient-Occident
GDA	Migrant Solidarity Network
GADEM	The Anti-racist Group for the Support and Defence of Foreigners and Migrants
HCA	Helsinki Citizens Assembly
HRDF	Human Resource Development Foundation
IOM	International Organization for Migration
LFIP	Law on Foreigners and International Protection
MAD	Moroccan Dirham
Mazlumder	Association of Human Rights and Solidarity for Oppressed People

MSF	Doctors Without Borders
MÜLTECİ-DER	Association for Solidarity with Refugees
NGO	Non-governmental organization
ODT	Democratic Organizations of Labour
ODT-IT	ODT-immigrant workers
OMDH	Moroccan Organization for Human Rights
POS	Political opportunity structures
RA	Readmission Agreement
SAFS	Social Assistance and Solidarity Foundations
TNP	Turkish National Police
TRY	New Turkish Lira
TOHAV	Foundation for Society and Legal Studies
UJRT	Union of the Young Refugees in Turkey
UNHCR	United Nations High Commission for Refugees

1   Introduction

André,[1] a 42-year-old migrant, originally from Cameroon, came to Morocco in May 2011, with the intention of going to Europe. After entering through Oujda, André spent several months in Tangier and in the forest near Ceuta and made several attempts to cross:

> When you make several attempts and when it does not work, you need to reflect on it [...] I have attempted several times in Tangier, several times in Ceuta. It did not work [...] We could organize among ourselves, buy a zodiac and [make an]attempt. [...] I told myself, I need to change, I would not say change tactic [sic], but my idea to go to Europe. I have decided that I can make my life here and in 2013, the King has given his discourse for the integration.

Since the summer of 2012, André has been involved with a migrants' solidarity association. The association was founded to raise awareness about racist attacks in poor neighbourhoods of Rabat. André has been doing voluntary jobs in collaboration with Moroccan associations and has actively worked to raise awareness about migrants' demands for rights and for regularization. While still dreaming of going to Europe, he is himself awaiting regularization.

Harun left Afghanistan in October 2009, at the age of 17, together with a cousin and two friends from his village. He planned to join his elder brother, who was living and working in Istanbul with other men from the village. After a three-week stay in Iran, they found a smuggler to take them to Istanbul. After crossing the border on foot, the smuggler took them to the United Nations High Commission for Refugees (UNHCR) office in the Iranian border city of Van. Harun went to the office to register without any knowledge of the asylum process in Turkey. 'I wanted to come to Istanbul, did not want to stay there. I never told them this.' During the application process, he explained that he wanted to go to Turkey to work and never mentioned his relatives in Istanbul or the smuggler. After leaving Van, Harun did not follow his asylum file. He arrived in Istanbul and settled in the flat shared by his brother and other single Afghan men. In the last three years, he has been living and working in Istanbul, moving from one workplace to another: 'Then, back in 2009, the work was scarce in Turkey.

---

1   All names are pseudonyms, unless indicated otherwise.

I had no jobs for the first two months. Then, I went to work in leather. [...] When the leather season was over, I left the job and went into the bag atelier.' He was later joined by his family members, who also crossed the border without documents. As of August 2013, the family had a pending residence permit application via their relatives, who were among Afghan nationals settled and naturalized in Turkey in the early 1980s.

Juxtaposing the stories of André and Harun illustrates the fragmented and dangerous journeys migrants must endure because of the existence of borders. The conditions of both men's journeys to the 'West' are similar, in the sense that they risked their lives crossing borders, getting help from smugglers, and facing the threat of detention and deportation, all in order to generate better opportunities in life. While there is a growing literature on borders and border crossings, this study is about the experiences of settlement beyond the borders of the European Union (EU). In addition to discussing the changing political environment, this book sheds light on how irregular migrants' 'uncertain legal status' (Menjívar 2006) within the national territories in which they reside is the result of law, practiced, and negotiated by the state, by civil society actors, and by migrants themselves. I incorporate migrant perspectives to help us grasp the processes that led André to become a political activist for migrant rights in Rabat, and Harun a textile worker in the informal sector in Istanbul. Interestingly, both have prospects for legalizing their 'illegal' status, but through different means.

The research questions reflect the multiple levels of analysis I embraced in addressing the question of migration governance at the periphery of the EU and irregular migrants' access to rights:

- How have changing policies and practices regarding the rights of irregular migrants produced migrant illegality in Turkey and Morocco as de facto immigration contexts?
- How do migrants experience their illegality and negotiate their presence in society in general, and their access to rights and legal status in particular?
- Under what circumstances do irregular migrants mobilize to claim their rights and legal status?

Through analysis of two country cases, this book contributes to the broader conceptual puzzle of how people in highly precarious positions, in terms of their relations to state authority, seek legitimacy. More specifically, my comparative inquiry aims to reveal the conditions under which irregular migrants in new immigration contexts may or may not seek 'political recognition', i.e. formal recognition of their presence and rights by authorities (Menjívar and Coutin 2014). I explore how this quest for recognition is interlinked with control mechanisms or, more generally, forms of governance of irregular migration that shape migrant illegality.

Amid growing concerns about irregular migration within the context of declining economic growth and the securitization of immigration, the developed world has adopted a more restrictive approach towards immigration and asylum. Particularly in the European migration system, emerging norms of EU migration controls have led to the expansion of security measures at the external borders of the EU. This research has primarily been motivated by the conviction that it is critical to explore what is happening beyond EU borders in terms of 'the production of migrant illegality' and 'migrants' access to rights'. This study not only conceptualizes irregular migration in the Mediterranean as an externalized EU border problem, but also looks at the different ways in which irregular migration becomes an issue of governance at the periphery of the EU. It is necessary for research to explore the implications of the increasing calls to halt irregular crossings at EU borders for the wider region, particularly for the people who suffer from policies and practices aimed at curtailing mobility into the EU. Especially in the context of current fatalities at the borders of the EU, the book provides a perspective on the conditions that have precipitated and, arguably, intensified the widely used notion of 'crisis' since the summer 2015. It does so by exploring what preceded the current 'migration governance crisis' at the external and internal borders of the EU.

I use the concept of 'governance' to refer to a multiplicity of actors and to policies as processes rather than end products. The term indicates that the focus is 'on processes of rule and not only on institutions' or on formal rules, but also on informal practices (Lemke 2007: 53). The term, as I use it, also refers to the fact that, in the realm of international migration, decisions and practices are contested by a variety of state and non-state actors; consequently, governments are not the only rule-making authorities (Betts 2011: 4). Meanwhile, the distribution of power and resources among these actors is unequal (Grugel and Piper 2011). The research suggests that changing migration policies, and their enforcement in Turkey and Morocco, have given rise to distinct forms of governance. Existing research has explored changes in the legal framework and the emergence of rudimentary immigration regimes in both Turkey and Morocco (Elmadmad 2011; Kirişci 2009). Little has been written, however, on how migrants themselves are influenced by changing policies and practices and how these practices are negotiated on the ground.

As two countries at the periphery of the EU, Turkey and Morocco have been subjected to the externalization of EU migration policies. In this context, a growing body of literature on EU migration controls, particularly on critical border studies, has focused on the external borders of the EU (Wunderlich 2010; Carling 2007; Collyer 2007; Mountz and Loyd 2014; Tsianos and Karakayali 2010; Pallister-Wilkins 2015). Since the early 2000s, Turkey and Morocco have increasingly been hosting immigrants who are either on their way to Europe, or who have crossed borders to look for opportunities to work, study and/or settle in relatively more developed countries in the region (İçduygu and Yükseker 2012; De Haas 2014). Despite this general observation on changing mobility patterns, less research has looked at the incorporation experiences that migrants and asylum seekers[2] have before reaching Europe (Collyer 2007; Suter 2012; Danış, Taraghi and Pérouse 2009). Even less research has explored the link between emerging forms of governance of irregular migration at the periphery of Europe and migrants' experiences of informal incorporation from a comparative perspective.

This book aims to address how legal frameworks produce migrant illegality in new immigration contexts, in which international politics applies pressure in order to govern unauthorized human mobility. This study analyses the production of illegality through emerging immigration policies and practices from a comparative perspective. In fact, comparative studies on migrant illegality are rare and rather new (Garcés-Mascareñas 2012; Lentin and Moreo 2015). Furthermore, few studies frame migrant illegality within an international context, in which illegality has resulted from interacting control and border regimes (Menjívar 2014). Given the recent changes to migration policies within the EU and new restrictions on mobility along EU borders, the book promises to explore how migrant illegality has been translated into these rather marginal spaces of immigration, beyond these borders, into what I refer to as 'new countries of immigration'. Morocco and Turkey, where immigration has only recently become a subject of governance, have been subjected to geopolitical pressures to stop irregular border crossings into the EU; they provide underexplored ground for re-thinking the processes through which migrant illegality has been produced, experienced, negotiated, and contested. To fill this gap, this book looks at how migrant illegality

---

2   While the book does not directly deal with asylum and refugee issues, as it is a specific area of international law, references are given to asylum issues especially when the issues pertaining to asylum and irregular migration are intermingled.

influences migrants' participation in economic, social, and political life, as well as how migrants challenge their 'illegal' legal status at the individual and communal levels.

By focusing on Turkey and Morocco as new immigration countries, the research brings together two levels of analysis; institutional, policy-oriented analyses on the impact of the external dimensions of EU migration control policies, one the one hand, and sociological analysis on migrant experiences of uncertain legal status on the other. The book addresses the missing link between migration governance and migrants' incorporation at the periphery of the EU in order to understand how irregular migrants seek legitimacy, while policies make them illegal.

This introductory chapter provides the overall conceptual frame for the following chapters of the book and details the methodological approach. The first part of the chapter reviews analytical tools to understand the processes through which irregular migrants are rendered illegal and subject to state controls; it looks at different ways in which irregular migrants participate in socio-economic life and negotiate their presence within economic, political, and legal structures despite their illegality. The second part elaborates on the methodological approach, where I discuss the logic of a comparative research design, the multi-layered data collection process, and the challenges of conducting fieldwork in two different contexts, the ethical issues emerging from my fieldwork experience.

1.1   Researching irregular migration as 'migrant illegality'

The term irregular migration generally refers to the presence of migrants in a given territory without authorization by the sovereign state. Irregular migration is more complex than crossing borders without the necessary documents. An immigrant with genuine entry documents, such as a tourist visa, could be living and/or working within the country with no legal status. An immigrant who is staying in a country legally with a residence permit may be considered an irregular worker if he/she is working without the necessary permits or beyond the authorized hours. An irregular migrant can also be a former asylum seeker whose application for refugee status was rejected. Despite the categories of legal and illegal fixed by law, people with no status may acquire a legal status, just as legal entrants or legal workers may fall into irregularity (Cvajner and Sciortino 2010: 214; Villegas 2014). Given the permeability between the categories of irregular migration and asylum and the malfunctioning of the asylum system, migrants may fluctuate between different legal and policy categories such as transit migrant, irregular migrant, or asylum seeker (Collyer and De Haas, 2012).

Considering this legal complexity, terms such as 'irregular' (with no regular/legal status), 'undocumented' (without the appropriate papers), and 'unauthorized' (without legal permission for entry, stay, or work) migration are used interchangeably to denote various facets of the wider phenomenon. Scholars are widely critical of the use of the term 'illegal migrant' or 'illegal migration', based on the simple notion that a person cannot be illegal (Van Meeteren, 2014: 18). The term 'illegal' reproduces state categories, portraying migrants as scapegoats rather than highlighting policies constructing migrants as 'illegal subjects'. This study uses the term 'migrant illegality' purposefully to centralize migrants' experience of lack of status and to reveal the meanings attached to the lack of status by different actors.

'Migrant illegality' as the central concept of my inquiry relies on Willen's conceptualization of the term: 'first, as a form of juridical status; second, as a socio-political condition; and third, as a mode of being-in-the-world' (Willen 2007a: 8). Following this tripartite definition, the research deals with three bodies of literature informing irregular migration research in general, and migrant illegality research in particular, to solve the puzzle of irregular migrants' access to rights and legal status. These include socio-legal studies on the legal production of migrant illegality, sociological research on irregular migrants' subordinate participation in society, and migrant political agency and other ways-of-being. The latter includes social movements literature that particularly focuses on cases of migrant mobilization despite their lack of political recognition.

How migrant illegality as juridical status is produced

The emergence of irregular migration, including transit migration as one form of mobility unauthorized by states, cannot be explained purely by the failure of migration governance or by a simple mismatch between socio-economic conditions in the sending areas that push people to emigrate and the receiving capacity of more developed regions (Cvajner and Sciortino 2010: 394). Irregular migration is a by-product of immigration policies rather than a gap between policies and their outcomes. The very existence of migration policies produces migrant illegality: 'There can be no illegal immigration without immigration policy, and thus the definition of those who are deemed to be "illegal", "irregular", "*sans papiers*", or "undocumented" shifts with the nature of immigration policy' (Samers 2004: 28). While most scholars agree that eliminating irregular migration is not a feasible goal, the socio-legal approach goes further to suggest that 'the law, thus creates the very subjects, on the surface, it seeks to bar' (Garcés-Mascareñas 2012: 31; see also, De Genova 2005; Coutin 2003; Calavita 2005).

The production of migrant illegality has been sustained through certain tactics of governmentality (De Genova 2004: 165; Willen 2007a: 13). These tactics range from deploying statistics/estimations of the presence of unauthorized non-citizens within the national territory to framing the phenomenon in particular ways representing irregular migrants as villains. Politically, reducing irregular migration to a technicality of numbers (of arrests, deportations) and to security budgets may serve to represent the issue within the sphere of national security and criminality. The convergence of immigration law with anti-terrorism and criminal laws reinforces the image of irregular migrants as a security threat to the nation and the social order. The criminalization of migration may go as far as classifying 'migration as a crime, penalization of humanitarian aid, criminalization of undocumented work' (Estévez 2012: 176). At times, irregular migration is equated with particular spaces or types of law-breaking, such as illegal border crossings or with particular ethnic groups of migrants.

Giving the impossibility of the absolute elimination of undocumented migration through deportation or detention, 'migrant deportability' does not necessarily mean actual exclusion, but implies its possibility. Practices of deportation differ in space and time. There are indeed 'geographies of deportation' (Garcés-Mascareñas 2012; Peutz and De Genova 2009). From a theoretical perspective, the threat of deportation functions as a disciplinary mechanism over migrants (De Genova 2004; Chauvin and Garcés-Mascareñas 2014: 423). Deportability makes migrants docile subjects who refrain from confrontation in the labour market as well as in social life. This process typically results in the economic marginalization of irregular migrants and reinforces their political exclusion.

In new immigration countries, those who would otherwise be called tourists and passengers are turned into illegal subjects as a result of the recent introduction of immigration laws and relatively stricter external and internal control measures that have been introduced due to external pressure. Furthermore, legal and administrative infrastructures and non-state actors were not prepared for this change and did not know how to deal with the new role of the country as a context of transit and immigration. Transposing the concept of the 'production of migrant illegality' onto the contexts under examination would thus require accounting for the national legal framework as well as the international context, imposing 'the gradual implementation of a system of migration management' (Samers

2004: 43) both within the EU and at its periphery. Hence, focusing on the periphery of Europe, I not only explore the production of illegality within the nation-state context, but also situate it within the broader context of the 'international production of migrant illegality'.

Relying on socio-legal studies on the legal production of migrant illegality, I transpose the question of the production of migrant illegality as a 'juridical status' onto new immigration countries where migrant illegality has resulted from external border relations. The EU has had a significant impact on both Morocco's and Turkey's immigration policies, hence the governance of irregular migration. I suggest that irregular migration has become an issue of governance in Turkey and Morocco in the last decade. In these contexts, state policies are shaped through the interaction of external pressures, i.e. the EU immigration regime and domestic dynamics. In other words, the interaction between EU and domestic factors have produced these transit spaces, which are unique spaces giving rise to particular forms of the production of migrant illegality.

Irregular migrants and subordinate incorporation

The literature on incorporation emphasizes that it is a process of inclusion into social life even in the absence of recognition from the state (Cvajner and Sciortino 2010: 398; De Genova 2004: 171). The divergence between law as written and law as practiced enables the presence of irregular migrants in formal and informal structures in society, otherwise known as 'semi-autonomous social spheres' (Moore 1973). Different terminology, such as 'legitimate presence' (Coutin 2003), 'liminality' (Menjívar 2006: 1003), inclusion into 'foggy social structures' (Bommes and Sciortino 2011), 'inclusion at a higher price' (Cvajner and Sciortino 2010: 400), 'subordinate incorporation' (Chauvin and Garcés-Mascareñas 2014), and 'integration in limbo', referring in particular to the case of transit spaces, (Danış, Taraghi and Pérouse 2009), has been proposed to explain this process. The book uses the term subordinate incorporation or informal incorporation to refer to the various processes that migrants such as André or Harun participate in, despite not being full members of society and in the absence of formal procedures.

As articulated in socio-legal studies, it is the law itself that produces 'illegality', which undermines the human rights of migrants and reinforces their vulnerable position in society (De Genova 2004; Calavita 2005). Here, one needs to take into account social as well as legal meanings of migrant illegality. In this sense, migrant illegality as a socio-political condition is shaped by discourses, institutional practices, and day-to-day interactions between migrants and state as well as non-state actors (Willen 2007a; Bommes and Sciortino 2011; Villegas 2014: 278). Research has underscored tensions between legal, institutional mechanisms excluding migrants without legal status from the political community, and migrants' de facto presence in the labour market, within welfare arrangements, and, at times, in political movements.

One important mechanism of what might be called informal incorporation stems from the gap between written laws and their implementation; in other words, the distinction between legal and social meanings of irregular migration (Bommes and Sciortino 2011: 217). The production of migrant illegality can take different meanings from one context to another, from one immigrant group to another. In the eyes of implementers, and in the eyes of migrants alike, there is a hierarchy of illegalities whereby some forms of irregular migration are considered more illegal, and the presence of some migrants is perceived as 'legitimate' regardless of their legal status (Kubal 2013). Coutin articulates, '[...] both the people being defined and the people doing the defining can influence the definitions produced, thus cumulatively "creating" law, in an informal sense of the term' (1998: 903). Thus, the process of 'cumulative creation of law' underscores that the law is re-formulated at the level of implementation, and this enables migrants to re-shape the categories they are put into. Therefore, looking at the everyday implementation of immigration law in various legal and socio-economic spheres, where legality is re-defined and re-produced, is equally important for revealing patterns in the governance of irregular migration as well as migrants' experiences of it (Coutin 1998, 2011; Kubal 2013). Hence, it is necessary to consider migrants' own experiences of inclusion and exclusion in depth to reveal 'local configurations of "migrant illegality"' (Willen 2007b: 3).

Discourses of control do not always coincide with actual practices that are often selective and arbitrary (De Genova 2002: 436). In spite of legal restrictions on entry and stay of migrants, states may largely tolerate the existence of irregular migrants within their territory. According to Amaya-Castro, weak illegality regimes occur, even in states with strong administrative capacities, when the number of those without legal status is perceived to be insignificant or other issues are deemed more important (2011: 142). It may also be the case that irregular migrants are tolerated because states benefit from their presence or prefer not to invest in the high administrative or financial cost of deportations. In this sense, no policy is also a form of governance whereby states refrain from taking responsibility for migrants' rights and protection simply by turning a blind eye to their existence, either by not regulating migration at all or by not implementing formal regulations. Chapter 4, for instance, talks about the urban labour market in Istanbul. Conversely, migrants' sense of illegality and deportability can further be reinforced through state practices, such as push-backs before migrants and potential asylum seekers can enter the country, frequent and unpredictable document checks, police raids in migrant neighbourhoods and workplaces, unlawful detention, and deportations (Galvin 2014). What Amaya-Castro (2011) would call 'strong illegality regimes' may also result in measures that breach irregular migrants' human rights recognized by national and international law. In such contexts, in which unlawful deportation practices are widespread and officials on the ground are resistant to granting status and rights to migrants, the possession of legal status may fall short of protecting migrants. What is even more striking than the suspension of law (in contexts in which laws are easily suspended) is the arbitrary implementation of law and the unpredictability of its outcome. This research contributes by revealing patterns in the arbitrary implementation of the law, looking at the governance and migrants' incorporation experience in contexts that are less constrained by liberal democratic norms.

What is called subordinate incorporation widely refers to the labour market conditions that incorporate migrants (Calavita 2005; Garcés-Mascareñas 2012). Studies have shown that the reproduction of the category of irregular migrant may serve the purpose of producing cheap labour for the economy (Calavita 2005). Therefore, several cases discussed in the literature focus more on labour demands. As implied above, the production of illegality in this research has been an outcome of external pressure that has occurred in the absence of, or regardless of, the state's explicit demands for labour. In other words, using sociological research on irregular migrants' subordinate forms of participation in society, the research explores how this external border closure interacts with labour market conditions in so-called transit spaces.

This process of subordinate inclusion is most visible in, but not limited to, migrants' participation in the labour market, where migrants gain a level of legitimacy through their economic participation in society, even when they lack a legal status. The general observation is that once irregular migrants are in the territory, they are incorporated into society through the informal labour market, but may also benefit from welfare institutions such as schools and hospitals through forged or genuine documents, become clients of humanitarian support, and participate in advocacy networks through (ethnic or religious) community-based mobilization (Cvajner and Sciortino

2010: 400; Chauvin and Garcés-Mascareñas 2012: 242). Comparative research may contribute to this body of literature by exploring processes leading to different styles of migrant incorporation. In other words, more empirical evidence is needed to theorize how contextual factors at international, national, and local levels impact 'migrants' individual and collective experiences of being-in-the-world' (Willen 2007a: 13).

A widely considered economic consequence of irregular migration is the fact that migrants' deportability renders them more vulnerable to exploitation in the labour market, especially in countries and specific sectors that are characterized by widespread informality (De Genova 2002: 439; Calavita 2005; Ahmad 2008; Villages 2014; Bloch and McKay 2016). The precarious work and exploitation it entails can be a form of migrant incorporation into social and economic life alongside other underprivileged segments of society such as unskilled legal migrants, ethnic minorities, and other underclass groups within urban economies. The informal economy constitutes one important mechanism of inclusion for irregular migrants as well as a potential way out of their illegality. Several studies have shown the implications of the absence of legal status with respect to precarious forms of labour market participation and irregular migrants' right to stay.

Labour market participation provides legitimacy to migrants' presence as subjects who contribute to the economy and thus deserve a legal status (Chauvin and Garcés-Mascareñas 2014). Regularization campaigns that offer the possibility for 'ex post legal inclusion' (Finotelli 2011: 205) aim at reducing the presence of irregular migrants by giving them legal status. Ironically, such campaigns require migrants' illegal presence to gain legal recognition (Coutin, 1998: 916-7). Garcés-Mascareñas' critique further emphasizes that, as a result of the legal changes in 2001 in Spain, 'work and not residence became the *sine qua non* condition for staying legal' (Garcés-Mascareñas 2012: 190). With reference to neoliberal citizenship, where the latter is conceptualized as an earned status, incorporation into the labour market has been perceived as grounds for legal incorporation. In other words, it is not necessarily the fear of deportability, but the prospect of being regularized, through work but also through other means, that becomes a disciplining factor for migrants and impacts their incorporation styles (Chauvin and Garcés-Mascareñas 2012; 2014). The expansion of trade unions' membership bases to include the (undocumented) migrant labour force provides another form of semi-formal incorporation of irregular migrants and may even provide migrants with a way out of irregularity. Meanwhile, there has been less research into the conditions under which labour market participation underpins migrants' quest for rights (Barron et al. 2011, 2016). Similarly, we also know less about the alternative ways that migrants without legal status may still claim legitimacy in the absence of labour market opportunities.

In addition to the economic sphere, migrant illegality has also been negotiated through formal institutions. As a consequence of the lack of legal status, public services constitute one of the main sites of exclusion for irregular migrants (Bloch and McKay 2016: 155-157). In contrast to this general view, previous research has revealed how undocumented migrants' rights have been extended through bureaucracy, before they have gained political recognition, in a process referred to as 'bureaucratic incorporation' (Marrow 2009) or 'bureaucratic sabotage' (Chauvin and Garcés-Mascareñas 2014: 424). This occurs in the daily acts, mostly by street-level bureaucrats (Lipsky 1980), who recognize migrants' legitimate right to access certain fundamental services. Without generalizing bureaucracy, Marrow (2009) suggests that, in the US context, most inclusionary practices towards newly arriving immigrants occur at the level of hospital emergency rooms and public elementary schools. Wilmes (2011: 130) uses the term 'useful illegality' to designate the provision of services to undocumented migrants under the rubric of a larger target group (people with no health insurance) in Germany. In Wilmes' analysis, providing healthcare to migrants without checking documents is illegal but useful, as it serves the general interest of public health and matches the ethical duty of treating a person in need of healthcare. Similarly, providing services to 'asylum seekers' in need of protection, regardless of whether they possess the necessary (asylum) papers has become the basis for most humanitarian organizations' legitimization of their services to irregular migrants (Coutin 1998: 908). The practices of bureaucratic incorporation show that migrants' access to institutions enabling fundamental rights may even constitute a mechanism of incorporation in contexts that are defined by economic and social exclusion.

There is documentation that suggests that bureaucratic incorporation in several contexts becomes possible when civil society intervenes. Humanitarian agencies are particularly interested in integrating those who cannot be easily absorbed by the labour market, such as pregnant women, women with small children, and elderly migrants. It is shown that when civil society provides services to irregular migrants, directly or indirectly, this substitutes for public welfare institutions and plays a role in reinforcing informal membership practices (Ambrosini 2013: 44; Taran and Geronimi 2003: 20). It is suggested that, by becoming beneficiaries of services, migrants are subjected to regularization from below (Nyers and Rygiel 2012). Therefore, the processes that enable access to fundamental rights demonstrate how illegality is negotiated on the ground, not only by migrants, but also by their pro-migrant rights allies. Further theoretical reflection is needed regarding the provision of public services to those who fall outside of formal membership, to contribute to the literature on 'street-level bureaucracy' in the context of migration controls (Van der Leun 2003: 28-29). Yet, a number of questions remain unanswered: Who benefits from this inclusion, which I will call 'street-level advocacy'? Who is left out? Under what configurations of illegality are irregular migrants conceptualized as legitimate clients/objects of humanitarian aid or rights-bearing political subjects?

Migrants as political actors?

In line with the literature on migrant illegality and migrant incorporation, I have so far suggested that migrant illegality is a product of immigration policies and is reversible on the ground through migrants becoming de facto members of society. The next section discusses how irregular migrants may contest and negotiate the stigma of illegality imposed upon them, and claim legal status through collective action and/or individual tactics. Social movements literature in relation to migrant illegality provides an opening for understanding the implications of concerted actions of irregular migrants for membership, even in less liberal contexts.

> Arendt does not show us the *sans papiers* only as victims, or as a disturbing signifier on the level of philosophical representation. By questioning state-centred thinking, the migrants appear also as political actors whose public appearance can be potentially explosive and liberating. (Krause 2008: 339, *emphasis original*)

Following Krause's reading of Arendt, the book conceptualizes the mobilization of migrants seeking political recognition as a form of incorporation into society. In this light, this section treats irregularity as a potentially reversible status. From this perspective, I discuss individual tactics and strategies of mobilization at the communal level, looking at how migrants negotiate their irregular status within these formal, semi-formal, and informal institutions. Highlighting the contrast between André's and Harun's trajectories to legalize their status, I question the conditions under which migrants actively seek the 'right to have rights', to become political subjects, and those which lead them to opt out of formal membership (i.e. legal status or citizenship). This is a puzzle to be explored further in the empirical chapters.

Exclusion from the political community, the risk of deportation, hostile discourses, and low prospects of being regularized may deter migrants from making rights claims and lead them towards further invisibility to decrease risks, but this also potentially increases vulnerability. Meanwhile, restrictions on mobility across borders and non-citizen access to status and rights have been challenged from the grass roots (Nyers and Rygiel 2012: 7; Nicholls 2014). Paralleling the politicization of irregular migration, and immigration in general, mobilization for the rights of irregular migrants has gained momentum in the developed world in recent decades (Nicholls 2013; Tyler and Marciniak 2013). Notably, migrants themselves have become part of these movements, despite the high risks involved. (Raissiguier 2014; Nicholls 2014).

The literature on the immigrants' rights movement discusses reasons for mobilization, as well as its mechanisms in terms of repertoires of mobilization, internal organization, and coalitions with other movements (Chimienti 2011; Tyler and Marciniak 2013; Mc Nevin 2012). Repertoires of resistance range from migrants' active use of social media, raising awareness about the fight against racist violence, outing themselves in public, and declaring the legitimacy of their presence (McNevin 2012: 177). Through these contestations, non-state actors, including migrants themselves, criticize the legitimate authority of the state by arguing that the deeds of the state vis-à-vis migrants may be within the law, but they conflict with other general principles, or by revealing the cases in which states have resorted to unlawful activities to get rid of irregular migrants (Kalir 2012: 48). Protests mainly problematize the taken-for-granted distinction between citizen and non-citizen (Tyler and Marciniak 2013: 147; McNevin 2006). Their presence within the territory and the simple claim that 'we are here' become legitimate grounds for migrants to ask for protection from violence and for their recognition and rights (Krause 2008: 342). Migrants' mobilization may occur in ethnicity-based solidarity groups, sectoral groups, or issue-based groups centred around the issue of lack of legal status or xenophobic violence/ discrimination (Nicholls 2013; McNevin 2006, 2012; Raissiguier 2014). As explained in Chapter 3, issue-based mobilization centred on deportations and racist violence has prevailed in the case of Morocco.

Studies have long employed the political opportunity structures (POS) approach, prioritizing the institutional environment to explain collective actions by migrants (Laubenthal 2007; Chimienti 2011; Nicholls 2013). Acknowledging the importance of pro-migrant actors and the importance of institutional factors, Però's and Solomos' (2010) review makes two substantive critiques that underscore my findings on irregular migrant mobilization in the case of Morocco. First, they argue that research using POS as the main explanatory factor has put insufficient emphasis on lived experiences as a key reason for migrants associating among themselves. They rightly point out other issues, such as political socialization, background, networks, and social capital of migrants, as key factors. Second, they explain that there is a need to include transnational opportunity structures in analyses of institutional contexts and pro-migrant rights alliances (Però and Solomos 2010: 9-10).

Migrants with no legal status need more resources than citizens and immigrants with legal status to participate in social life and to mobilize and advocate for their rights (Cvajner and Sciortino 2010). Undocumented migrants need the support of citizens to further their interests (Breyer and Dumitru 2007: 138), to recognize political opportunities available to them, and to provoke reactions from other actors in the field (Bröer and Duyvendak 2009). Indeed, political mobilization by migrants themselves and by pro-migrant activists go hand in hand; one important component of mobilization is the forging of 'unexpected alliances that migration creates' (Coutin 2011: 302). One emerging hypothesis from migrant mobilization literature, to be tested through comparative case analyses, is whether it is less likely for irregular migrants to mobilize among themselves without the support of a pro-migrant rights movement.

As articulated by Tyler and Marciniak (2013: 152), 'it is of critical importance that we examine the ways in which irregular migrants and their allies negotiate the contradictions, losses and gains of in/visibility in their interactions with sovereign power.' While existing research mostly analyses where immigrant subjects are politicized and actively seek recognition, cases of non-mobilization are equally important. Visibility and representation carry risks of exposure to state control (Tyler and Marciniak, 2013.), and therefore mobilization may not always be desirable for irregular migrants. Chimienti's (2011) comparative study analyses POSs for immigrants' rights movements in three European cities. Chimienti argues that not only restrictions, but also a shift in state practices from tolerance to restriction is a factor in migrants' mobilization and also influences pro-migrant rights actors. The case of Paris, where regularization campaigns and labour market opportunities have become increasingly exclusionary, is an example of mobilization that extends beyond ethnic ties around the issue of irregularity (2011: 1343). From a comparative perspective, migrants' mobilization is more scattered and more ethnically divided in the case of London, where illegality regimes generate interstices for tolerance and legitimacy. In the case of Copenhagen, invisibility and a lack of interest from non-governmental organizations (NGOs), which are focused on asylum-related issues rather than irregular migrants, are factors contributing to irregular migrants' lack of mobilization (Chimienti, 2011: 1348). While Chimienti's comparative lens is useful, my research goes one step further by exploring the link between mobilization, that is being politically active as a mode of being-in-the-world, and other incorporation styles, in relation to other aspects of migrant illegality as a juridical or socio-political condition. Irregular migrants may activate alternative 'social resources that compensate for the lack of inclusion in the political system' (Bommes and Sciortino 2011: 224-5). At this point, it is necessary to explore the manifestations of migrant illegality that lead irregular migrants to opt for or against the risks involved in mobilization.

Individual tactics

The tactics that migrants use to stay in the territory in the absence of political inclusion may or may not be directed at gaining formal recognition. Staying invisible but tolerated, in other words 'illegal but licit', may also be a useful survival strategy for migrants. As Coutin emphasizes, 'for some groups, the primary need is to avoid deportation not to seek for legal status' (1998: 905). Rights, or the possession of legal status, may not be a priority as long as the threat of deportation is not experienced daily. Furthermore, migrants aspiring to continue to other destinations, or perceiving their stay as temporary, may not feel an immediate need for recognition from the state. In other words, it might be in the interest of some irregular migrants to stay invisible and avoid state control.

To avoid the attention of authorities and the possibility of deportation, migrants avoid petty crimes and neighbourhood or workplace conflicts (Chauvin and Garcés-Mascareñas 2014: 426). Migrants also consciously choose not to send their children to school, avoid going to public hospitals unless absolutely necessary, and abstain from written communications because these are ways that they can be identified and targeted by the authorities (Breyer and Dumitru 2007: 139-140). At the same time, as theorized by Chauvin and Garcés-Mascareñas (2014), the term invisibility falls short in terms of depicting migrants' ways of being-in-the world. It is rare for irregular migrants to have absolutely no contact with public institutions and civil society organizations (CSOs) that provide welfare services and do advocacy work on their behalf; they are rarely fully undocumented. A considerable portion of irregular migrants (certainly legal entrants) holds passports, entry documents, and identity cards from their countries of origin. The possession of (the right) papers is crucial, especially in contexts of strong illegality regimes where deportation is a daily threat, and irregular migrants are perceived as a security threat. Research reveals that migrants constantly collect legitimate identification papers from their countries of residence, such as a municipality registration, driving licence, birth certificates for their children, asylum application documents, etc. Forged documents may also ensure a legal presence, especially in contexts where administrative procedures do not work properly (Sadiq 2008).

Staying docile in the shadow economy and possessing genuine or forged identification papers (not necessarily the proper ones) allow migrants to stay under the radar until they have the opportunity to reverse their illegal status (Chauvin and Garcés-Mascareñas 2014: 411). Migrants may get opportunities to acquire legal status through their own efforts, for example by convincing employers to apply for necessary work permits, applying for student residence permits by enrolling in schools, or through marriage. When there is a prospect for regularization, migrants are especially active in negotiating their presence by being 'visible enough' without becoming 'too visible' (Chauvin and Garcés-Mascareñas 2012: 252).

Using this conceptual toolbox, the book will unpack the interconnection between immigration policies, migrant incorporation styles, and irregular migrants' tactics to access rights and legal status. Regarding migrant illegality as a 'way of being-in-the-world', I blend sociological literature on migrant incorporation into society and on contentious politics. The study questions the interactions between social and institutional mechanisms that give rise to very different styles of incorporation. As implied in the ethnographic vignettes juxtaposing the stories of Harun and of André, the book explores how migrants of irregular legal status in Morocco have managed to raise political demands for their entitlements to rights and legal status despite stigmatizing and hostile contexts. Conversely, it questions how irregular migrants in Turkey have become de facto members of society without political voices. By explicating the mechanisms of migrant incorporation styles, my empirical findings question if it is necessary for migrants to be political subjects in order to legitimize their presence. Furthermore, I question the extent to which migrants' political claims for legal status depends on their presence in the labour market.

1.2   Researching migrant illegality in new immigration countries

Contributing to existing literature on migrant illegality and on irregular migrant incorporation, the research transfers these discussions to new immigration countries, where migrant illegality is a relatively recent phenomenon, resulting from the international situation, while not necessarily tied to labour market demands. Through the empirical discussion in the two contexts, I focus on the interrelatedness of the production of migrant illegality, the production of a quiescent labour force, and mechanisms of migrant activism. The book aims to inform more general discussions and theories of how and through which mechanisms marginalized and legally excluded groups gain legitimacy.

This study uses a comparative research design to shed light on the processes that give rise to different incorporation styles in different contexts, intending to contribute to the emerging literature and theorization on forms of migrant illegality. The case selection is based on the two countries' similar emigration histories, directed towards Europe since the second half of the twentieth century, and on their similar geographical locations at the periphery of Europe, a factor that makes them de facto lands of immigration. I use the phrase 'de facto lands of immigration' together with 'new immigration countries' to underscore that these countries have become transit and destination points without their explicit political will or economic need for immigration.

First, I have looked at the migration regimes characterized by strict external controls and more or less rigid internal controls for curtailing irregular migration, considering their implications for the production of migrant illegality. De facto immigration contexts such as Turkey and Morocco, at the periphery of Europe, as well as Mexico in the North American context, are good examples for observing foreigners who were once considered licit despite lacking the necessary papers to stay, work in the country, or passing through the country. Furthermore, these contexts have become subject to governance since the 1990s. They not only cover a wide range of irregular migration, from overstaying one's visa to fraudulent entry, but there are also contexts in which foreigners in irregular situations are additionally categorized as 'transit', based on their alleged intention to leave for their final destinations. Hence, the category of 'transit' further complicates the production of migrant illegality and further excludes migrants without legal status from the political sphere of membership in the contexts under scrutiny. Therefore, researching irregular migration in contexts characterized with transit (im)mobility would require the analysis of the production of migrant illegality at an international level.

Second, I have conceptualized migrant incorporation styles as an outcome of interactions occurring through the legal production of migrant illegality, practices of deportability, social and economic structures in the receiving society, and the availability of an institutional context that is conducive to shaping and channelling rights claims. One implicit hypothesis in migrant illegality and incorporation literature is that the production of migrant illegality gives rise to a cheap labour force, readily exploitable in the labour market. In her comparative inquiry on the connection between market demands for cheap labour and rights constraints in Malaysia and Spain, Garcés-Mascareñas (2012: 31) suggests that whether the production of migrant illegality turns into the production of cheap, flexible labour is more of 'an empirical question than a starting point of inquiry.' This empirical question is even more open-ended in the comparison of Turkey and Morocco, as new immigration countries where the production of migrant illegality has resulted from the international contexts surrounding them, rather than an explicit demand and political will to receive migrants. Another related, open-ended question is whether informal incorporation into the market provides a source of legitimacy for irregular migrants' presence in the society and the extent to which labour force participation provides a basis for migrants' quest for legal status, insofar as it is *deserved* through one's contribution to the economy.

As underscored by the literature on the experience of illegality, irregular migrants actively participate in society in different ways; they negotiate their visibility in the public sphere (Willen 2007a), seek to legalize their status, and, at times, get mobilized and forge alliances to claim their rights to legitimately reside in the territory (Laubenthal 2007; Nicholls 2013). Research has indicated links between configurations of migrant illegality, irregular migrants' incorporation experiences, as well as their experiences of political mobilization (Willen 2007a, 2007b; Laubenthal 2007). However, more research and analytical reflection are needed on the conditions under which experiences of marginalization may or may not lead to mobilization. Such an approach would put migrant experiences at the centre of analyses without necessarily neglecting the political opportunities that are available to migrants or the roles played by pro-migrant rights allies.

At a theoretical level, the analysis contributes to the theorization of the link between the governance of irregular migration and migrants' incorporation, reflecting on the relationship between control and recognition: Does the quest for recognition by the authority necessarily imply the acceptance of control by the same authority? Or, is it possible that irregular migrants would seek recognition in response to the strict controls imposed upon them; in particular, socio-economic and institutional settings that push and pull them towards mobilization? The conceptualization of settings within which migrants are incorporated, as transit rather than destination, would impact the relationship between control and recognition.

1.3   Comparative research design and case selection

Comparative research design is the primary instrument used in this study to reveal mechanisms of irregular migrant incorporation in contexts that are subject to similar external pressures to control and manage irregular migration. As Theda Skocpol puts it, in her contribution to the Symposium on Comparative Politics, 'the purpose of comparison should be partly to explore and test hypotheses from a variety of theoretical perspectives and partly to notice and hypothesize about new causal regularities' (Kohli et al., 1995: 38). At the same time, comparative research designs entail episte-mological challenges. When compared to single-case analyses, comparative research lacks equal depth and thickness of understanding in the collection of data as well as in the presentation. In Sartori's words (1991: 253): '[in case studies] one knows more about less (in less extension). Conversely, compara-tive studies sacrifice understanding – and of context – to inclusiveness: one knows less about more.' Acknowledging the promises and limitations of comparative research design, this section looks at how the cases under scrutiny are comparative, how the data is collected, the challenges involved in conducting research in two field sites, as well as the ethical challenges involved in research with vulnerable populations.

Earlier research on irregular migration in the Mediterranean pointed to Turkey and Morocco as comparable sites for looking at the impact of external dimensions of EU policies (Fargues 2009; Scheel and Ratfisch 2013; Papadopoulos, Stephenson, and Tsianos 2008: 165). In terms of the generalizability of my findings, the analysis does not claim that Turkey and Morocco are representative of peripheral countries that are subject to external dimensions of the EU migration control regimes. However, the case selection is likely to reveal the differential impact of EU border measures on the two nation-state contexts most affected by these measures. Despite their differences in terms of the existence of colonial ties, the scale of their economies, state capacities, and colonial regimes, there are certain key factors that have enabled the comparison of the two nation-state contexts. Most notably, these include their similar migration histories, as migrant-sending regions to Europe, their similar geopolitical positions, and their relation to the EU. More specifically, they have common historical transition patterns that saw them change from countries sending labour migrants to Europe into lands of destinations (İçduygu and Kirişci 2009; de Haas 2014). Moreover, both countries receive similar types of flows in terms of transit migration, asylum, labour, student, and retirement migration, albeit from different source geographies, as explained in Chapter 2. Another basis for the comparison between Turkey and Morocco is their geographical similarity. Both countries are located at the tightly controlled gates of Fortress Europe, at both ends of the Mediterranean, which has been identified with irregular migratory flows since the early 1990s. Their geographical similarity also makes them similar in their position towards external aspects of EU migration policies. Turkey and Morocco have become subject to similar pressure to control their EU borders. In their comparative work on the role of UNHCR in Turkey and Morocco, Scheel and Ratfisch (2014: 927) highlighted the fact that in both contexts, 'migration has not been framed and treated as a "problem" that needs to be regulated until a short time ago.'

For a relevant analysis across cases, Landman (2003: 35) underscores that important concepts should be specific enough to measure what the research intends to measure in each case and general enough to cover all cases in question. The novelty and external character of the debate render the processes of the production of migrant illegality in the two contexts studied comparable. Terms such as irregular, illegal, and transit migration are borrowed from the EU policy agenda and evoke similar social phenomena and legal categories. Both cases commonly represent a particular interaction between the international and domestic contexts, leading to the emergence of irregular migration as a governance issue and rendering migrants illegal subjects before the law. In other words, 'state simplifications', to use James Scott's terminology (Kohli et al. 1995: 29) on the question of irregular migration, have emerged in comparable terms.

One direct implication of the new and external character of the issue has been the underdeveloped legal framework regarding international migration in general, irregular migration in particular. The legal frameworks on immigration in the two counties have gone through changes in the post-2000 period. Migration policies simultaneously represent a reaction to incoming flows of migrants and the external pressure to control these flows, with few concerns for migrant rights. In the cases under scrutiny, irregular migration emerged as a subject of governance in similar terms at around the same time. Given their changing roles from migrant-sending countries to countries that act as gateways that control irregular migration, to sites of immigration management, both countries are constrained in the process of stopping irregular mobility flows to the EU and respecting fundamental rights.

Along with the geographical, political, and historical aspects explained above, personal and practical reasons influenced the case selection. Being from Turkey and interested in irregular migration within Turkey has contributed to my focus. The selection of Morocco as a comparative case has arisen from my personal interests in the Mediterranean region. My fluency in French and already established relations with scholars working on Moroccan migration made Morocco a viable option for my comparative inquiry.

1.4   Data collection

In terms of conducting fieldwork, comparative research designs require dividing the fieldwork time rather than focusing on a single case. I collected the data on the case of Morocco over several visits. I divided my fieldwork time into three intense visits between April and October 2012, each lasting around three weeks. I paid two shorter follow-up visits in March and May 2014 in the aftermath of the reform initiative. The timeframe of the fieldwork in Turkey has been more flexible, as I reside in the country. I conducted the interviews between January 2012 and December 2013.

Dividing the fieldwork time brought advantages as well as disadvantages. Morocco was a new terrain of research for me, and it took time to become familiar with the migrant scene as well as to introduce myself to different actors. During some of the interactions, I regretted not staying in Rabat for longer periods to strengthen trust relations within migrant communities and activist networks and to better grasp the daily power relations in encounters with the state, as well as within the community. Aside from the practical reasons, dividing fieldwork time enabled me to travel back and forth, not only physically, but also mentally between data collection, analysis, and literature review. Data gathered and pre-analysed during initial visits affected my data collection strategies for later visits. Dividing the fieldwork time provided me with the necessary mindset to constantly compare and contrast the two cases. While conducting fieldwork and learning specific aspects of migrant illegality in Morocco, I always kept in mind the specificities of Turkey. One particular challenge was to keep the data collection process balanced in the two contexts. Differences in the contexts and in my subjective position as a researcher influenced my access to resources.

By means of qualitative methods, I have explored emerging forms of governance and modes of incorporation of irregular migrants in Turkey and Morocco between 2000 and 2014, a period when irregular migration became an issue of governance and academic research. The research methodology mainly borrows from political science, the sociology of migration, and socio-legal studies. Going beyond the dichotomy of studying up or studying down, parallel to other research on the subject of irregular migration (Van der Leun 2003; Tsianos and Karakayali 2010), I embraced the approach of studying through 'tracing policy connections between different organizational and everyday worlds' (Shore and Wright 2003: 11) by collecting data at various sub-national levels by triangulating perspectives of various state and non-state actors involved. I employed a three-layered comparative research design to trace differences in the mechanisms through which illegality is produced and irregular migrants participate in social, economic, and political life for each case in question. To this end, data is primarily generated through the analysis of legislative documents and interviews with stakeholders, including state officials, civil society actors, and migrants.

Legal documents

Analyses of legal documents provided necessary background on the legal conceptualization of irregular migration and the availability of certain procedural and fundamental rights to irregular migrants. As Shore and Wright (2003: 26) described, policy analyses are necessary to understand 'how policies work as instruments of governance, as ideological vehicles, as agents for constructing subjectivities and organizing people within systems of power and authority.' In both countries, I looked at the legislation on foreigners' entry, residence and works permits, acquisition of citizenship and asylum, and deportation procedures. The documents for analysis were selected in order to reflect the diversity of legal and illegal categories constructed by law to reveal the connection between control over irregular migration and recognition of migrant rights on paper. The focus of document analysis is on the legal construction of illegality and the rights that irregular migrants have on paper, as these countries are becoming countries of immigration with a gradual official acknowledgement of the changing mobility situation.

The access to official statistics was limited in both contexts, but particularly prevalent in Morocco. In Turkey, in theory, anyone is entitled to make inquiries and ask for official data. In practice, I did not always get positive responses to my inquiries, and the information received was not as detailed as requested. In the end, I was able to obtain statistics from institutions and from secondary literature, which gave an indication of irregular migration in both contexts, although the data gathered may not always be comparable.

Expert interviews with state officials and civil society actors

Given the focus of enquiry, there was an evident need to go beyond official state perspectives. In order to understand the functioning of laws,

I conducted expert interviews with state officials and representatives of international organizations and NGOs, with 22 institutions in Morocco and 17 institutions in Turkey. Informants included law makers and high-/mid-level bureaucrats dealing with issues of immigration, and representatives of international and national NGOs and inter-governmental organizations.[3]

Semi-structured interviews generally explored the activities of key institutions on immigration and asylum-related issues in the post-2000 period. Expert interviews intended to reveal the general framing of issues pertaining to immigration and to discern external and domestic dynamics leading to legal changes. Questions probed migrant profiles and the changing legal framework regarding migrants' access to rights and legal status. Informants were invited to reflect on the different categories emerging in law such as legal, irregular migrants, asylum seekers, and refugees. In Turkey, most of the interviews took place on the eve of the new Law on Foreigners and International Protection (LFIP) was introduced and came into force. Consequently, I asked informants explicit questions about their views on the new legislation and on their participation in the process of law making. In the case of Morocco, legal changes were initiated after the completion of the fieldwork. However, follow-up interviews in March and May 2014 not only complemented earlier interviews, but also enabled me to grasp the changing policy discourse. In both contexts, while some state officials would simply repeat what was written on paper as a validation of the official discourse, others provided insightful information on the functioning of laws, enabling me to have a better understanding of the discrepancy between written laws and practice. The insight gained from these interviews has been crucial in revealing and comparing the local and institutional dynamics in the implementation of laws in both countries. Interview findings are triangulated with observations in public meetings organized by state institutions and/or civil society. To complement interview data, especially in the case of lack of access to certain institutions, I analysed institutional documents (press releases, reports, etc.) and media outlets, including public statements by government officials.

Regarding the selection of institutions interviewed, the primary criterion was explicit interest and expertise in the area of immigration and asylum. For instance, I did not approach trade unions in Turkey, because irregular migration has not been on the agenda of trade unions in Turkey, unlike in Morocco where they were approached for interviews. Similarly, migrant organizations have been either formal ethnic associations established by

3   See Tables 1-2-4-5 for more information on stakeholders interviewed.

migrants who have acquired citizenship and/or ethnicity-based informal solidarity networks. Including their members as informants in Turkey would require doing the same in Morocco, which would mean interviewing members from every single formal and informal ethnicity-/nationality-based migrant association. Instead, I limited my inquiry to associations making political demands on behalf of irregular migrants in general, rather than for particular ethnic groups. The visibility and accessibility of migrant organizations in Morocco and the invisibility of those in Turkey shaped the list of informants in both contexts.

My outsider position in Morocco and my insider position in Turkey impacted the data collection process. Differences were marked regarding the institutions I could access for interviews. I was able to conduct interviews in general police departments and in the Ministry of the Interior in Turkey. In Turkey, I tried to use the advantage of being an insider. Certain interviews were possible because of my professional connections, whereas for others, I conducted interviews without any intermediaries at the institution. Approaching the Ministry of the Interior was out of the question in Morocco. None of the people I met could, or were willing to, connect me with a person in the Ministry of the Interior or Foreign Affairs, and my formal attempts were inconclusive. However, the bureaucrats responsible for the Migration Directorate in Morocco were more visible in the national media than their counterparts in Turkey. By scanning news outlets in Francophone Moroccan media – I do not have the language skills to scan the Arabophone media – I was able to document official statements since the department's establishment in 2003. Additionally, my participation in policy meetings organized by state institutions and civil society press releases proved very fruitful for my data collection in Morocco. These were productive for grasping different arguments, meeting potential informants, catching up with others already interviewed, and even asking follow-up questions outside of the formal interview setting.

Migrant interviews

In order to reveal migrants' experiences of legal status and the ways in which they negotiate their access to rights, interviews with migrants of different legal status, i.e. undocumented, (rejected) asylum seekers, and overstayers – mostly persons moving between legality and illegality – were conducted in each country. In parallel with the expert interviews, the migrant interviews probed four major issues around migrant experiences of illegality: (i) controls by authorities; (ii) labour market situation; (iii) access to fundamental rights; and (iv) political mobilization and other tactics to reverse illegality or to negotiate visibility. Interviews revealed migrants' own accounts of their illegality, their experiences of deportation and settlement, as well as the social and legal mechanisms available to them to gain access to rights and legal status.

I had to be careful and strategic in building trust relations with (potential) informants and in remunerating both gatekeepers and informants. With the help of other researchers or migrants that I met through these researchers, I started paying regular visits to neighbourhoods where migrants reside, work, do business, perform religious activities, call their families, hang out, etc. These visits enabled me to make ample observations and engage in small talk with migrants and locals. I had the chance to hire foreign students as research assistants in both Rabat and Istanbul. These students live in migrant neighbourhoods and/or are familiar with different migrant communities. Gatekeepers were particularly helpful in neighbourhoods that can be unsafe for a young woman, especially after dark. However, the presence of an intermediary also had the potential to cause informants to self-censor. Once I familiarized myself with neighbourhoods and initiated personal relations with people living in those neighbourhoods, I preferred to conduct interviews one-to-one, if there was no translation needed and if respondents were comfortable talking to me.

The interviewees were reached with the help of several gatekeepers and through the personal connections I developed during my visits to neighbourhoods, intending to get a purposeful sample that reflected the diversity of immigration experiences in both settings. Brief encounters were not always fruitful for arranging formal interviews, especially in Istanbul where migrants were busier with work (in comparison to Morocco) and were reluctant to talk to strangers. Conversely, the migrants I met, especially in Rabat, were willing to talk even after initial encounters. While the problems that arose in each context were different, the issue of access was present in both.

The snowball technique, which is recognized as an appropriate way to access hard to reach groups, was used in a limited fashion. In certain cases, one key informant enabled me to interview several others from his/her own community; however, it was not the case that each informant referred me to new ones. I had to initiate several starting points to achieve diversity among informants in terms of country of origin, legal status, demographic factors, and tightness of their connection to institutions. Needless to say, my main aim was to interview migrants without authorization to reside or work in the country, i.e. illegal entrants, overstayers, informally working residence permit holders, and rejected asylum seekers. Indeed, it has proved difficult to distinguish whether one is a potential asylum seeker, an asylum applicant, an economic migrant with no papers, or a residence permit holder without doing in-depth interviews. Interviewees were informed about the research and they participated on a voluntary basis. Rather than financial remuneration, I provided some of them with necessities. For example, I invited them to eat with me, and I gave them small gifts (food, fruits, desserts, milk or toys for their children, chocolates on special days, etc.), especially when they invited me to their homes.

In total, I interviewed 35 migrants (16 women and 22 men) in Morocco and 30 migrants in Turkey (16 women and 14 men).[4] I acknowledge that the purposeful sample was heterogeneous in terms of education, reasons for migration, migration aspirations, family status, and so on. In this sense, the research refrains from reproducing categories of needy irregular migrants who are low on social and economic capital (Cvajner and Sciortino 2010: 394). I also tried to go beyond the stereotype of single young men associated with transit migration. Migrant narratives have been triangulated with other sources of information. Each interview lasted between 40 minutes and two hours, dependent on how much time the migrants had available. In some cases, I had the chance to conduct several interviews. I conducted interviews in French, English, and Turkish. In Morocco, all the migrants encountered spoke either English or French. In Turkey, I asked my gatekeepers to act as translator in seven interviews because informants were either unable to speak Turkish, or felt more comfortable expressing themselves in their native language despite their understanding of Turkish. To ensure continuity in the narrative, some of the interviews are quoted more often in the empirical chapters. This is not to prioritize experiences of some over others, but rather because they articulate a common pattern more concisely than others. Moreover, some experiences extracted from interviews and observations are summarized without direct quotations.

Although the research does not claim to be a fully-fledged ethnography, because of the limited time spent in each research site, I incorporated observation as an ethnographic method into my research design. To complement interviews, I made observations in social milieus frequented by migrants, such as neighbourhoods, call centres, internet cafes, churches and gatherings during religious holidays, and home visits. In addition, whenever possible, I engaged in small talk with locals in neighbourhoods where migrants reside to grasp local perceptions of the presence of foreigners.

---

4   See Tables 3 and 6 for information on basic information on migrants interviewed in the two contexts.

Ethical issues and negotiating resources

Ethical measures have been taken to protect human subjects directly or indirectly participating in the research. As required by the Koç University Ethical Board, the informed oral consent of all informants was gained beforehand, and interviews were tape-recorded only when they consented. I had to make strategic decisions on the issue of recording. Rather than recording the interviews with state officials, I preferred to take extensive notes during most interviews both in Turkey and Morocco. The issue of recording was much less problematic with civil society representatives. Interviews took place in a friendly atmosphere, even when I asked critical questions regarding Turkish NGOs' neglect of the question of irregular migration or regarding tense relations with the Moroccan and Turkish states. I always made sure that it was possible for me to stop recording if they wanted to provide some information off the record. I ensured that our conversations ended in a friendly manner by thanking the interviewee and turning off the recorder when I invited the individuals to reflect upon my research and my questions. I prefer not to use the name of stakeholders interviewed because some of the statements are sensitive. When necessary, I indicate the institutional affiliation of the person, especially when it is important to note the type of institution that has generated the particular information, rather than the particular person that I interviewed from that institution.

Regarding interviews with migrants, the interviewee would decide whether or not to record the conversation. Every time I felt any hesitance from the side of the informant, I put the recorder away and preferred to take extensive notes instead of making a recording. In contexts in which there are power hierarchies between the researcher and the researched, I made it clear that informants were free to refuse to answer my questions or stop the interview. I kept the structure of the interviews as loose as possible, especially at the beginning of interviews when I collected migration stories. I waited to ask more specific questions at the conclusion of the interview. I did my best to show my appreciation for the information they provided, even though I sometimes had the feeling that some aspects of the stories were not true. I tried to probe inconsistencies. I made notes of these points to return to in the following meeting, if possible, or as question marks for my analysis.

The recordings, their transcriptions, and my interview notes were kept securely. The material was made anonymous, coded, and managed using N-Vivo software. While transcribing interviews, I made clear notes on what issues had not been raised by the informants, as well as those subjects they preferred to bring up without prompting. During coding, I generated explicit memos on my perceptions of what was willingly or reluctantly told to me. I did not use direct quotes from unrecorded interviews unless my notes were clear enough that the statement was a direct quotation. The anonymity of the interviewees was ensured by keeping any possible information that may identify interviewees out of the analysis. Keeping informants anonymous is a crucial component of ethics in this research as the individuals involved are either state officials, i.e. people in power positions, or migrants, i.e. people in vulnerable situations.

My subjectivity in the field had a direct impact on my access to different sources of information in the two field sites. During my fieldwork in Morocco, the extent to which my gender and ethnicity shaped my experience in the field became clear. Everyone was interested in the fact that I was from Turkey, and this was definitely more interesting than if I were American or European. At the level of institutions, people were asking questions about life and the situation in Turkey, as Turkish TV serials are shown on Moroccan channels, and Turkey had become a popular destination for the Moroccan middle class. I always felt that I was expected to look more modest than Western female researchers, as I was from a Muslim country. It was comfortable for me to wear loose clothes and no make-up in order to diminish looks from Moroccan men and migrants. As an outsider, as a young woman from a Muslim country, interested in Morocco, I was welcomed in different venues. I was able to meet some officials because I was a foreigner who had travelled to their country for a limited period. Being a 'white' woman from Turkey, researching Africans in Morocco, migrants in Morocco were much more willing to talk to me than those in Turkey. Immigrants that I interviewed and met also asked me a lot of questions about Turkey. Some were willing to stay in touch. I could sense that they were considering Turkey as a future destination. I also faced ethical dilemmas as I was seen as a person capable of helping irregular migrants get documents such as asylum papers, residence permits, or visas. I had to clarify that I was not connected to an authority that could grant them papers, but I was open to helping with paperwork such as translation, writing petitions, or dealing with bureaucracy.

My discussions with Moroccan researchers in the field gave me the impression that sub-Saharan migrants are more inclined to complain about the situation in Morocco to a foreigner than to a Moroccan. Parallel with this observation, I feared that migrants in Istanbul may not be opening up to me, whom they consider an insider, as much as they would to a foreign researcher. To overcome this bias, I crosschecked my findings with other Turkish and non-Turkish researchers who have conducted research in the same neighbourhoods.

Conducting interviews with migrants in Istanbul is challenging without intermediaries. Because of the long work hours of the majority of inform-ants, most interviews took place during weekends. I showed respect and appreciation for being able to conduct several interviews during migrants' very limited leisure time. Despite the challenges of access, being physically present in Istanbul enabled me to have frequent face-to-face and phone contact with the informants and build trust relations. Frequent contact has been crucial to understanding how migrants change legal status and gradually develop strategies to participate in socio-economic life, get legal status, or arrange their future journeys, and how these strategies might fail.

Conversely, I was not physically present in Morocco after October 2012. Indeed, I left the country when migrant activism and demands for the regu-larization of undocumented migrants were at their peak and when there was no apparent prospect for improvement. Between this time and the launch of the regularization campaign in November 2013, the internet provided me with the opportunity to continue collecting data on how irregular migrants in Morocco represent their situation and demands using different media outlets including Facebook and local, national, and international media. In both contexts, being Facebook friends with (potential) informants initially helped me to build trust relations because informants became familiar with me (my physical appearance, my work, my civil status, etc.). At times, social media enabled me to follow the mobility of individuals across borders.

1.5   Mapping the book

Chapter 1 framed the conceptual and methodological tools that I used in my study. I sketched out the theoretical implications of the production of migrant illegality and migrants' incorporation for new immigration countries. The chapter raises theoretical and empirical questions to be resolved in later chapters: How do new laws and institutions, practices of state and non-state actors, as well as socio-economic structures shape migrants' strategies to access rights and legal status? The second part of the chapter elaborated on the methodological approach.

In line with the main theoretical, methodological, and empirical motiva-tions of my research, the rest of the book is structured in five chapters. Chap-ter 2 explores how the international context contributes to the production of migrant illegality in new immigration countries and also reflects on domestic factors. The chapter describes the external and internal dynamics through which irregular migration has become a policy concern. The impact of the international context, mainly EU policies leading to the emergence of transit spaces, is taken as a distinctive aspect of the production of migrant illegality in the contexts in question. The emergence of Morocco and Turkey as transit spaces, the EU's impact on the emergence of immigration and border policies, and the political and institutional context within which policies and practices towards irregular migration have taken place are explained from a comparative perspective. Thus, this chapter contributes through its focus on the international and national dynamics that impact the production of migrant illegality, offering insight on the implications of this interaction from a comparative perspective.

Chapters 3 and 4 focus on practices that relate to the production of migrant illegality and migrants' incorporation experiences in Morocco and Turkey, respectively, in the post-2000 period, introducing perspectives from migrants and civil society actors. Detailed analyses are provided on the practices of producing (reinforcing, tolerating) migrant illegality and on migrants' access to the right to stay and to services. I discuss how migrants' experiences of incorporation are shaped by practices on the ground and policies as well as the structure of the labour market and the interventions of non-state actors. I suggest that individual and communal strategies are available for migrants to get access to rights and legal status. The chapters provide an empirical answer to the sociological question that the book addresses: 'How do migrants seek legitimacy and access rights and legal status, as nation-state policies and practices make them illegal?' Chapters 3 and 4 are structured as mirror chapters to enable interested readers to cross-read sub-sections. I explain each country case separately to enable the reader to follow the interaction among the production of migrant illegality, migrants' experiences of incorporation, and their strategies for accessing rights and legal status in each country case.

Building on the insights of Chapters 3 and 4, Chapter 5 is a systematic comparison of the production of migrant illegality and irregular migrants' experiences of incorporation at the periphery of EU borders. The chapter argues that the production of migrant illegality arguably gave rise to different forms of incorporation despite the similar international context that led to the production of migrant illegality at the edge of European borders. Thus, Chapter 5 refines the findings of my research by explaining the prevailing forms of economic, social, political, and legal incorporation in both contexts. After sketching the major differences in migrants' experiences of incorporation (without overlooking similarities), Chapter 6 (the concluding chapter) refers back to theoretical and empirical puzzles that were introduced in Chapter 1 on different aspects of migrant illegality. As discussed in the concluding chapter, the research findings are prone to generating hypotheses for further studies on migrant illegality and on the incorporation of irregular migrants in new as well as old immigration countries.

2 The production of migrant illegality

International and domestic dynamics in a comparison

> The EU's external policy is producing a new geography of remote control,
> which extends beyond carrier sanctions and placing customs officials in
> third country airports. (Samers 2004: 40)

> [Migrant illegality] is a product of converging global, regional, and
> national factors. (Willen 2007a: 27)

Introduction

The evolution of international mobility patterns in the contexts of Turkey
and Morocco has been analysed, as country cases, from historical, sociological, and political perspectives (İçduygu 2006, 2007; Kirişci 2008; De Haas
2007, 2014; Castles 2007; Iskander 2010). Both Turkey and Morocco have a
considerable number of citizens living abroad, predominantly in European
countries. In both contexts, migration policy has mainly referred to emigration policy. Due to the EU's interest in remote controls to prevent irregular
migration (Samers 2004), Morocco and Turkey initially assumed the role of
transit spaces and have only recently become new immigration countries
under the pressure of EU border policies (Düvell and Vollmer 2009; Scheel
and Ratfisch 2014).

The novel and external character of the emergence of irregular migration
as a subject of governance makes the two country cases comparable with
regards to the production of migrant illegality. They are suitable cases to
explore the production of illegality in relation to the international context.
Both contexts have conventionally been studied as emigration countries
in the literature, yet irregular migration has only emerged in the last decade as a category of governance and a subject of academic studies. These
are the contexts in which the production of migrant illegality is a recent
phenomenon. In other words, migrant illegality is a new process developed
throughout the late 1990s and 2000s through the diffusion of norms, laws,
and institutions, mainly as a result of tightening EU border policies. Using
the insight of existing research, this chapter puts this transformation from
emigration to new immigration countries in a comparative perspective
along the main problematic of my research, namely the framing of irregular

migration as a policy issue and irregular migrants as illegal, deportable subjects.

This chapter looks at the interaction between the EU migration regime and emerging immigration policy in peripheral contexts. Irregular migration policies in Morocco and Turkey have been the subject of analyses in the context of external dimensions of EU migration control policies (Alami M'chichi 2006; Wunderlich 2010; İçduygu 2007; Ozcurumez and Şenses 2011). One line of research has analysed the export of EU migration control and migration management techniques without necessarily delving into the implications of this expansion for the production of migrant illegality (Boswell 2003; Samers, 2004). Another approach has called for a focus on borders sites and on the migrant body as a vulnerable subject of these externalization policies (Mountz and Loyd 2014; Tsianos and Karakayali 2010). Papadopoulos, Stephenson, and Tsianos have drawn attention to the 'productivity of the European migration and border regime' at the periphery of Europe (2008: 165) and present the examples of Moroccan-Spanish and Greek-Turkish borders as sites of this production. However, few studies concentrate on the interaction between external dimensions of EU migration policies and migrant illegality beyond EU borders. As Menjívar puts it, 'the construction of immigrant "illegality" (De Genova 2002; Menjívar and Kanstroom 2014) is no longer confined to the territorial borders of the receiving country; it is a process that starts before immigrants arrive at the physical border, in transit areas and, in some cases, even at the point of departure' (2014: 363). Scholars have underscored that migrants are subject to the interacting migration regimes long before they reach EU shores (Karakayali and Rigo 2009: 125; Brigden and Mainwaring 2016). Based on this observation, this chapter aims to show how the restrictions imposed by the EU migration regime have influenced national policies and the kind of migrant illegality this interaction has produced.

The chapter first briefly summarizes the evolution of mobility patterns in Morocco and Turkey to provide a comparative lens on immigration patterns. Second, I explore the international context in terms of the external dimensions of EU migration control policies that triggered legal and institutional change, starting in the early 2000s until the end of 2014. With reference to migrant illegality literature, I call this process 'the international production of migrant illegality.' As explained in Section 2.2, the externalization of EU migration policies in the post-2000s period manifests itself through certain policy tools of governance that are common in both contexts. Among these, I focus on increasing investment in border infrastructures, cooperation agreements (such as readmission agreements (RAs)) with the EU, and the intensification of activities by international and intergovernmental organizations, especially UNHCR and the International Organization for Migration (IOM). The third and fourth part of the chapter discuss how this external production of migrant illegality has been translated in the domestic sphere in each case, giving rise to the politicization of irregular migration in the two countries studied. Note that the EU policies and responses to these policies are interlinked. The distinction between external and internal/domestic aspects of the governance of irregular migration is rather analytical. Descriptions of processes of the legal and institutional changes provided in this chapter will be used in the following chapters to explain causal connections among the production of migrant illegality, migrants' experiences of incorporation, and migrants' access to rights and legal status.

2.1   Becoming lands of destination

A closer look at mobility patterns reveals that immigration is part of the national history of Morocco and Turkey. Immigration was initially a policy matter in the colonial context in Morocco and in the context of nation-building in the case of Turkey. Colonial relations have been influential in shaping the mobility patterns in Morocco in the pre-1960 period (Berriane et al. 2010: 18). Until the 1960s, Morocco was a land of immigration (for the French, but also for its southern neighbours). Throughout the twentieth century, Turkey has been a land of immigration for Muslim and Turkic groups from its wider region, but these arriving groups are perceived as natural citizens rather than foreigners. According to the 1934 Settlement Law, immigrants are defined as those people of Turkish descent and culture who come to settle in Turkey. The policies shaped around this logic reveal continuity in the sense that, even today, some groups or individuals can more easily access legal residence and citizenship on the basis that they are of Turkic descent (see Danış and Parla 2009). The policies developed to respond to the arrival of 'ethnic kins' were widely challenged by the arrival of 'real foreigners' coming to Turkey to work and/or continue on to Europe in the post-1990 period (Erder 2007).

Initiated by bilateral labour agreements signed in the 1960s, Turkey and Morocco's emigration histories have emerged as directed towards Europe. The numbers of workers originating from Turkey or Morocco who live in different European countries have significantly increased since then. Despite the changing migration regime in Europe, which put an end to the mass recruitment of migrant labour, emigration to European countries continued through family reunification and later through family formation (De Haas 2009; İçduygu and Sert 2009). After the 1980s, irregular migration (for both Turkey and Morocco) and asylum (for Turkey) have become major types of flows to Europe. The introduction of a visa requirement for Moroccan nationals to enter Spain (1991) and Italy (1990) and for Turkish nationals to enter France (1980), Germany (1981), the UK (1989), and the Netherlands (1996) reinforced irregular migration from Turkey and Morocco (De Haas 2014; Doğan and Genç 2014: 230).

Despite this change, there were still less barriers to travel to Europe than today (Collyer and De Haas 2012: 471). Irregular migration was not yet a hot topic connected to security and social cohesion issues. Migrants without the necessary papers as well as asylum seekers were mostly seen as spontaneous guest workers in the epistemological and political terrain of migration (Karakayali and Rigo 2010: 130). It was possible for migrants and asylum seekers from Morocco and Turkey to legalize their status after their arrival (Collyer 2007: 670-671). Consequently, asylum seekers from Turkey – mostly of Kurdish origin – joined the labour force in Western Europe and could eventually become legal residents. Moroccans in irregular situations in France and in southern Europe benefited greatly from regularization campaigns (Garcés-Mascareñas 2012: 158). The change in out-migration patterns reveals that irregular border crossings were initially an issue pertaining to emigration. Turkish and Moroccan nationals are still represented among nationals crossing borders without valid documents in recent Frontex reports (Frontex 2014). This point is beyond the focus of my research, but it is suffice to say that irregular border crossing as a form of mobility for Turkish and Moroccan nationals has not disappeared but declined. The decline is more significant in the Turkish case. However, in both contexts, the attention of the EU and national policymakers has shifted to third-country nationals.

Conventionally source countries, since the 1990s, both Turkey and Morocco have had to assume new roles in the European migration system as transit and eventually as immigration countries (Fargues 2009; De Haas 2007; Kimball 2007). Within the context of globalization and continuing relations with European migration regimes, Turkey and Morocco had already started to receive immigration from their wider region. As a result of the EU's interest in preventing irregular migration, the presence of migrants seeking clandestine entry into Europe – given the decreasing opportunities for legal entry – has become more visible and more subject to state regulations. Such regulations also affected migrants in Turkey and Morocco who had purposes other than moving into Europe, such as working, studying, and seeking refuge.

Morocco predominantly receives migrants and asylum seekers from African countries, such as Nigeria, Mali, Senegal, DRC Congo, and Sierra Leone (Fargues 2009; Mghari 2009; Berriane et al. 2010; AMERM 2008). Given the growing obstacles to crossing into Europe, migrants from sub-Saharan countries (commonly called sub-Saharans) have become increasingly visible in urban centres such as Casablanca, Rabat, Tangier, and, more recently, Fez (De Haas 2007; Berriane and Agerdal 2008). While statistics and official data on immigration into Morocco are far from complete, the estimates of the number of sub-Saharan irregular migrants between 2000 and 2010 ranged from 10,000 to 20,000 (see Khachani 2011: 4). By the end of the regularization programme in December 2014, over 27,600 migrants with an irregular status had applied for the regularization scheme, and a further 26,000 cases were apprehended at the border, providing another source for estimating the volume of irregular migration in the country (Ministry in Charge of Moroccans Abroad and Migration Affairs 2016: 79-83). In 2013, over 4,300 people entered the enclaves of Ceuta and Melilla through clandestine means (APDHA 2014: 47). Although higher than previous years, the number is still much lower than the number of clandestine migrants within Morocco estimated based on the number of apprehended cases. Until 2012, illegal entries into the enclaves did not exceed 2,000 per year (APDHA 2010: 10). The discrepancy between those entering the Spanish enclaves and those remaining in Morocco without status indirectly indicates that Morocco has become a land of (forced) settlement for thousands on their way to Europe, along with those arriving in Morocco to work or study. The number of asylum seekers and recognized refugees has remained relatively modest, around 4,500 people in September 2015, as Morocco does not share borders with conflict-generating regions in the African continent.

Besides irregular migrants, Morocco receives a significant number of international students from sub-Saharan Africa, some of whom are sponsored by the Moroccan government (Berriane 2009). Another trend in Morocco is the settlement of Europeans who buy properties in big cities. This population movement is considered insignificant from the policy perspective, as settlers are not conceptualized as a threat to national security and people do not self-identify as migrants, but rather as expatriates. The number of legal residents (a total of 86,206 as of September 2014) and irregular migrants in Morocco constitutes less than 1 per cent of the total population and is by no means comparable to the number of emigrants originating from Morocco. Despite the relatively low number of incoming migrants with or without legal status, immigration into Morocco has been the subject of increasing academic and policy-oriented research. Most of the existing research concentrates on the most salient figure of the 'illegal migrant' in Morocco, i.e. undocumented sub-Saharans allegedly on their way to Europe.

Since the 1980s, the geographical situation of Turkey, coupled with relatively liberal visa policies, has enabled different forms of undocumented entry and stay by foreign nationals (İçduygu and Yükseker 2012). Similar to Morocco, the data for comprehending the volume of irregular migration in Turkey is limited. Looking at apprehended cases is an inadequate tool, yet the most convenient for estimating the volume of irregular migration in Turkey. The number of migrants apprehended by security forces has rocketed from around 11,000 to nearly 100,000 in 2000. Since 2000, there has been a declining trend in this number to nearly 40,000 in 2013, rising again to nearly 59,000 by the end of 2014. In the same period, asylum applications have significantly increased from a few thousand in 2005 to over 34,000 applicants in 2014.[1] Since the 1980s, the country has experienced sizable asylum flows from Iran, Iraq, Afghanistan, and, to a lesser extent, African countries. These numbers do not include Syrian refugees under temporary protection since the breakout of the Syrian conflict in 2011.[2] The Syrian case illustrates the external and asylum-related character of migration management and, also indirectly, migrant illegality issues in Turkey. Turkey adopted an open border policy with Syria, enabling the settlement of Syrians fleeing the conflict in refugee camps close to the Syrian-Turkish border (Ihlamur-Öner 2013). Syrian refugees, initially settled as 'guests', were later granted 'temporary protection' status in October 2014. In other words, in principle, Syrians are excluded from the UNHCR status determination process.

Ahmet İçduygu has analysed irregular migration in Turkey under three broad categories: i) Transit migrants who intend to cross to the EU through Turkey and usually enter the country without proper documents with the paid help of smugglers (see İçduygu 2006, 2007). Transit migrants, allegedly on their way to Europe, tend to come from southwestern Asian countries such as Afghanistan, Pakistan, Iran, Iraq, and lately Syria (İçduygu and Yükseker 2012); ii) Irregular labour migrants who typically enter Turkey with a valid visa and work in the informal economy without valid documents.

1   See Statistics by the Directorate General of Migration Management (DGMM) Retrieved 06.11.2016, from http://www.goc.gov.tr/icerik/migration-statistics_915_1024
2   See ibid. By the end of 2014, the official number of registered Syrian refugees in Turkey was 1.5 million, then rocketed to 2.8 million by the end of 2016.

When compared to Morocco, the economic aspect of irregular migration is much more salient in Turkey. The country hosts economic migrants from countries of the former Soviet Union including the Turkic Republics, Ukraine, Moldova, and Armenia; iii) Asylum seekers who originate from the same countries as migrants put into the transit category, and who enter the country without proper documents. Some of them are admitted into the asylum system in Turkey and are awaiting recognition by UNHCR and eventual resettlement in third countries. As İçduygu also acknowledges, this typology rarely fits individual migrants' trajectories. There is a thin line between asylum seekers and irregular migrants, as most migrants move between legality and illegality as well as between transit and settlement in Turkey. Meanwhile, it is a useful typology of how different groups of migrants fall into illegality, hence become deportable, as further explained in Chapter 4. Regardless of their aspirations to go to Europe or acquire legal status, migrants find employment opportunities in sectors such as domestic work, sex work, entertainment, textiles, construction, and tourism. Migrants are mostly concentrated in big cities, predominantly Istanbul.

Turkey's immigration and asylum toll is much larger when compared to Morocco's, and migrant profiles are more diverse in terms of country of origin. One should also note that Turkey's overall population (81 million) is 2.5 times larger than Morocco's (31 million). As depicted in Table 7 in the Annex, the numbers and profiles are not easily comparable. What is comparable, however, is the emergence of political debate. Immigration and, more specifically, irregular migration have emerged as policy issues although, in demographic terms, Turkey and Morocco are not yet ageing societies in need of migrant labour. There has been no political will to receive immigration. The governance of irregular migration has entered into the political agenda as an outcome of 'migration diplomacy' with the EU (Natter 2014; İçduygu and Üstübici 2014).

2.2   The international context in the production of illegality

Restrictions have characterized the European asylum and migration regimes since at least the mid-1980s. In this context, concepts such as 'first country of asylum', 'irregular secondary movement' (Oelgemöller 2011: 415), 'transit migrants', or 'stranded migrants' (Dowd 2008) have appeared to refer to what I call the 'international production of migrant illegality' and its transposition to peripheral countries. As both border and visa regulations are highly developed policy domains in the EU (Samers 2004: 34), cooperation with transit countries on the issue of irregular border crossings has been one of Union's primary policy tools (Boswell 2003). Morocco and Turkey have similar geographical positions and close relations to the EU, both of which, among other important factors, lead to a particular production of illegality. Moreover, due to their location and geopolitical relations to the EU, the shift in the EU's policy priorities have deeply affected policies and practices relating to irregular migration in Turkey and Morocco. These changes, also depicted in Annex 3, have turned the countries from spaces of 'transit' into 'anti-transit' areas where alleged transit migrants are stopped and controlled.

In the African context, Morocco is a major example of the securitization of migration (Belguendouz 2009; Elmadmad 2007). Morocco is only fourteen kilometres away from mainland Spain, separated by the Strait of Gibraltar. Furthermore, in the north, Morocco neighbours two Spanish enclaves on the African continent: Ceuta and Melilla. The enclaves are around 100 kilometres from the Algerian-Moroccan border, where most migrants cross without legal papers. The Canary Islands, one of the outermost regions of the EU, are reachable from the southwestern coast. Moreover, the southern borders of Morocco are not clearly defined because of the political dispute over Western Sahara. They are relatively more permeable to intra-African mobility due to lax border and visa regimes. With the growth of Spain's economy throughout the 1980s, the income differences on both sides of the border have become drastic and, along with irregular migration from other parts of the world, have triggered irregular migration from Morocco to Spain. The proximity to EU borders and the political conviction to stop transit migration make Moroccan-Spanish borders a primary subject of migration diplomacy among Morocco, Spain, and the EU.

Turkey is at the crossroads of Asia, Europe, and Africa. Geopolitically, it is located between asylum seeker- and migrant-generating regions and European destinations. Most of the Greek islands in the Aegean Sea are only a few miles away from Turkey's western coast. The border between Turkey and Greece has been identified by Frontex as 'one of the areas with the highest number of detections of illegal border-crossing along the external border' (Frontex 2012: 4-5). As in the case of Morocco, the geographical proximity to the EU made Turkey's western border subject to securitization. In the east and south, however, Turkey shares land borders with Georgia, Armenia, Iran, Iraq, and Syria. These borders are permeable to the mobility of goods and humans due to lax visa regimes, challenging geographies, regional conflicts, and historically established economic and social relations between both sides of the nation-state borderlands.

The 1980s was a period when both Turkey and Morocco neglected the phenomenon of their own nationals and increasingly third-country nationals crossing through their territory into the EU. The official negligence continued until the countries were identified as transit zones by the EU. Assuming the role of a transit country implies subscribing to certain techniques of governance that render migrant populations deportable. These techniques range from increasing border controls through to financial and technical assistance by the EU (Samers 2004: 38-9) and readmission agreements (RAs). RAs are seen as a key instrument for preventing irregular border crossings into the EU. These agreements set the procedures for identification and return of persons 'who have been found illegally entering, being present in, or residing in the Requesting State.'[3] Other techniques include cooperation agreements with EU agencies, member states as well as increasing activities by international organizations.

Morocco's migration diplomacy

Immigration first became a subject in international relations and, later, in the context of political isolation, in internal politics in Morocco. During the 1980s and 1990s, Morocco had been politically isolated and had limited cooperation with the EU and tense relations with North African and sub-Saharan countries (Natter 2014). Morocco left the African Union in 1984 due to strained relations with its eastern neighbour, Algeria, over sovereignty in Western Sahara the eastern border with Algeria, located near the city of Oujda, has been closed since 1994 (Perrin 2011: 9). Morocco's application to become an EU member state in 1987 was rejected. Throughout the 1990s, tensions prevailed in Moroccan-Spanish relations because of Morocco's alleged tolerance of illegal migration within its territory.

Morocco has been identified as a transit space since the late 1990s, although it is known that Spain has been receiving irregular migration flows since the 1980s. The Association Agreement between the EU and Morocco was concluded at the end of 1995 and came into force in 2000. In this document, both parties agreed to initiate cooperation on illegal immigration and the conditions governing the return of irregular immigrants (DEMIG 2015). More concretely, the 1999 Action Plan proposed by the High-Level Working Group on Asylum and Migration was an early document urging a

---

3   See for instance the Readmission Agreement between Turkey and the EU signed on 16.12.2013, Article 1, Retrieved 06.02.2015 from http://www.resmigazete.gov.tr/eskiler/2014/08/20140802-1-1.pdf

halt to irregular border crossings through Morocco. The plan required the government to conclude readmission agreements that would also cover third-country nationals and to adopt visa requirements for West African nationals (JAI 75 AG 30 1999: 15 cited in Natter 2014: 18). As discussed in Natter's (2014) analysis, while Morocco rejected this plan imposed by the EU, the country strategically used the EU's interest in irregular migration to improve its relations with the Union and started to engage in migration-related diplomacy.

Throughout the 1990s, Morocco's relations with Spain and the EU were characterized by friction over irregular migration. As a new destination country for Moroccans as well as for third-country nationals travelling through Morocco, Spain has been a key player in Morocco-EU relations (Wunderlich 2010: 263). The visa requirement increased entry through the Spanish enclaves of Ceuta and Melilla, which hold special status outside of EU Schengen borders and where Moroccans are permitted to enter with a valid passport for a maximum of 24 hours (Zapata-Barrero and Witte 2007: 86). To prevent illegal entries, the Spanish government started to build fences and walls around the enclaves in 1993. According to Zapata-Barrero and Witte (2007: 86), this was the first step towards the securitization of the southern Spanish borders as a whole. According to Lutterbeck (2006), this has resulted in a change in the locality of transit migration in Morocco. As measures were taken around Gibraltar, Ceuta, and Melilla, the irregular routes shifted towards the coasts near the Canary Islands. In response, the integrated border surveillance system known as SIVE was established in 2002 and 'had reached full coverage of more than 500 km of Spain's south coast and was due to extend to the Canary Islands by the end of 2007' (Collyer 2007: 672). By 2001, due to the dynamic and, relatively, rapidly changing nature of migration routes, the Spanish and Moroccan governments were forced to confront the challenge of cooperation.

On the part of Morocco, this led to the introduction of Law 02/03 and the establishment of a department uniquely responsible for irregular migration (Valluy 2007). As part of the cooperation, since the early 2000s, Morocco has received technical and financial assistance to enhance its border control system (Wunderlich 2010). The EU made up to 70 million Euros (Nieselt 2014: 13) available within the context of cooperation measures designed to help Mediterranean non-member countries. In 2006, Morocco, in collaboration with France and Spain, hosted the first Euro-African Ministerial Conference on Migration and Development. The aim of the conference was to establish a global dialogue on migration. This initiative is an example of Morocco's ambition to be a regional leader. Morocco was seeking a credible regional leadership role 'as a migration manager at an international level by playing the role of a lobbyist of Mediterranean and African concerns' (Wolff 2008: 263).

Another instrument of external governance by the EU member states and a means of cooperation with third countries are the readmission agreements (Cassarino 2007). Since the 1990s, Morocco has signed readmission agreements with individual European countries including France (1993, 2001), Germany (1998), Italy (1998, 1999), and Portugal (1999). These agreements entailed the readmission of nationals but excluded third-country nationals (MPC 2013: 178). Despite prevailing undocumented border crossings from Morocco to Spain, the readmission agreement between Morocco and Spain signed in 1992 was only ratified by Morocco in 2012 (Cherti and Grant 2013: 14), and it was never fully implemented because of Moroccan reluctance to admit nationals and, in particular, third-country nationals (Cassarino 2007:183; Garcés-Mascareñas 2012: 170).

Another phase in these negotiations was the Mobility Partnership Agreement signed with the EU and six EU member states in June 2013. Along with initiatives to ensure the legal migration of Moroccan nationals, the agreement aims at enhanced cooperation 'to prevent and combat illegal migration,' as part of 'the exemplary partnership which has linked Morocco and the EU for several decades.'[4] In this regard, one of the key aims of the document, as articulated in Article 13, was the completion of a readmission agreement between Morocco and the EU, with provisions relating to third-country nationals, which had long been at the negotiation phase. Articles 28 and 35 of the partnership document envision EU assistance for the introduction of a new asylum and international protection system in Morocco and for an improved legal framework concerning various categories of migrants. In this sense, the document signalled changes in the Moroccan immigration policies that would be initiated in 2013.

In the context of closer cooperation with the EU and Spain, international organizations working on migration, mainly the IOM and the UNHCR, signed formal agreements with the Moroccan state. The agreement with the IOM was signed in 2006 and entailed 'efficient management' of the migratory question in Morocco and a budget to finance voluntary return (Valluy 2007: 6). The EU and member states such as France, Belgium, the

4   See, Joint declaration establishing a Mobility Partnership between the Kingdom of Morocco and the European Union and its Member States, Retrieved 23.02.2015 from http://ec.europa.eu/dgs/home-affairs/what-is-new/news/news/2013/docs/20130607_declaration_conjointe-maroc_eu_version_3_6_13_en.pdf

Netherlands, and Spain are major donors to IOM projects in Morocco.[5] While Morocco ratified the 1951 Geneva Convention in 1956, the representation of UNHCR in Morocco was merely symbolic until 2004. After a period of de facto functioning of UNCHR's office in Rabat between 2004 and 2007, the Headquarter Agreement with UNHCR was signed in July 2007. The agreements legitimitized the UNHCR refugee status determination by granting a residence permit to those with refugee status (DEMIG 2015). In practice, the card provided by UNHCR protects refugees from deportation but does not give access to residence permits and work permits in the country (Elmadmad 2011:.4). Valluy (2007) identified the two main factors behind the country agreement between UNHCR and Morocco. One was the changing priorities in the EU external policies in terms of impeding secondary movements of refugees from transit spaces. The second was the increasing number of applications to the UNHCR office from those unofficially settled in Rabat since 2004.

In the international context, EU demands have been influential in shaping the policies and practices of Morocco towards irregular border crossings of its own nationals as well as third-country nationals. The relations between Morocco and the EU and between Morocco and Spain are characterized by tensions as well as 'à la carte cooperation,' as discussed by Wunderlich (2010: 266). Immigration has become a permanent topic in foreign relations not only with Europe, but also with Morocco's southern neighbours, as EU policy demands have increased inner-African deportations and removals (Trauner and Deimel 2013). After the 2000s, EU-led international actors, mainly UNHCR and IOM, started to operate in Morocco and immigration has become a subject of governance par excellence. As will be explained in Section 2.3, the post-2003 period witnessed the institutionalization of immigration governance within the state, but the simultaneous emergence of myriad domestic actors shaped and re-shaped the practices around migrant illegality in Morocco.

Irregular migration in Turkey's long-standing EU accession

Drawing on the communication between the Intergovernmental Consultations on Migration, Asylum and Refugees, UNHCR, and Turkey, Oelgemöller (2011: 414-5) suggests that Turkey was the first country to be identified as a

---

5   See Morocco Country Profile, retrieved 06.02.2015 from http://www.iom.int/cms/en/ sites/iom/home/where-we-work/africa-and-the-middle-east/middle-east-and-north-africa/ morocco-1/country-profile.html

transit space, as early as 1987, for its role as a first asylum country for refugees fleeing conflicts in the region, such as the Iranian Revolution, the Iran-Iraq War, and the Gulf crisis. Meanwhile, Turkey's long-standing member status and its commitment to adopting the EU acquis have been major anchors for Turkey's cooperation with the EU on the issue of irregular migration. In this context, legal changes in the field of asylum and immigration in the post-2001 period are commonly called the Europeanization of migration and asylum policies in Turkey (Özgür and Özer 2010; İçduygu 2007; Ozcurumez and Şenses 2011). As in the case of Morocco, issues related to border controls, the resolution of a readmission agreement, and the increasing role played by international organizations in the context of the adoption of the EU acquis on migration and asylum, have been major milestones in the international production of migrant illegality in Turkey.

Administrative, financial, and technical support by the EU and member states significantly contributed to making irregular migration a subject of governance (Özgür and Özer 2010: 138-9). The National Action Plan for Asylum and Migration, adopted in March 2005, was a product of a twinning project with Denmark and the UK, conducted between March 2004 and March 2005. The Action Plan envisaged legislative and institutional changes to harmonize Turkey's asylum and migration legislation with that of the EU acquis. Officially starting with the 2003 Strategy Paper for the Protection of External Borders, border management issues have been on the agenda concurrently with membership talks, along with migration management and asylum issues. The framework of the Action Plan on Integrated Border Management, adopted in 2007, was initiated alongside another twinning project in collaboration with the UK and France. The EU funded a considerable portion of the budget to conduct these projects. For instance, the EU is estimated to have contributed nearly 22 million Euro (slightly more than 75 per cent of the total budget) for the execution of the second phase of the Action Plan on Integrated Border Management.[6] The EU's conditionality and financial and administrative support in border management issues highlight the novel and external character of the emergence of irregular migration policies in Turkey.

In the context of integrated border management, the EU expects Turkey and Greece to cooperate on matters related to border security. The readmission agreement between Greece and Turkey came into force in

---

6   See Standard Project Fiche, the Action Plan on Integrated Border Management-Phase 2, Retrieved 06.02.2015,from http://ec.europa.eu/enlargement/pdf/turkey/ipa/2008/tr080210_action_plan_on_ibm_phase_ii-revised_final_en.pdf

2002. Like the readmission agreement between Spain and Morocco, there have been severe problems of implementation, not least due to Turkey's reluctance to agree to readmit irregular migrants who allegedly crossed into Greece through Turkey (İçduygu 2011: 7). According to data compiled by İçduygu (2011: 7), Greece made 65,300 readmission claims to Turkey between 2002 and 2010. Of these, Turkey agreed to re-admit 10,124 persons, and only 2,425 readmissions actually occurred. Increasing cooperation with Frontex along maritime borders led to a decrease in interceptions on the sea borders between Turkey and Greece (Frontex 2012: 18), but also to a shift in clandestine routes towards the Evros region and at the Bulgarian-Turkish land border (Özgür and Özer 2010: 107-8). Despite its economic crisis, the Greek government, together with the EU, co-funded the construction of a double-fenced, 12.5-kilometre-long wall along the border, echoing Spain's erection of a wall around its enclaves in Northern Africa.[7] These measures, however, did not stop irregular border crossings and instead diverted smuggling routes and enhanced migrants' reliance on smuggling networks, raising the cost of border crossings. The situation along the EU-Turkish border closely affects migrants' experiences of incorporation within Turkey. Given the increasing costs and risks of crossing into the EU, transit migrants allegedly spend more time in urban centres in Turkey and seek ways to survive within given economic, legal, and social structures like those migrants categorized as asylum seekers and irregular economic migrants.

As another aspect of EU migration controls, after many years of negotiations, Turkey signed a readmission agreement with the EU in December 2013. The readmission concerns the nationals of EU member states and Turkey, plus third-country nationals and stateless persons who 'entered into, or stayed on, the territory of either sides directly arriving from the territory of the other side' (EC 2013a). Turkey signed the RA in exchange for the initiation of the EU-Turkey visa liberalization dialogue. The deal signed in March 2016 between Turkey and EU, envisages further curtailing and criminalizing of unauthorized border crossings and deportation of migrants and refugees back to Turkey in exchange for visa liberalization for nationals of Turkey. In other words, Turkish nationals' potential visa-free travel to European countries depends on Turkey's efforts to stop irregular migration into the EU. Interestingly, especially the RA, but also the Turkey-EU statement of

---

7   Plans for a wall on Greece's border with Turkey embarrass Brussels. *Guardian,* 11.01.2011. Retrieved 22.02.2015, from http://www.theguardian.com/world/2011/jan/11/greece-turkey-wall-immigration-stroobants

2016 was presented in the media as a technical commitment on the part of Turkey to open the borders of Europe for its own nationals. There was less discussion and almost no official statement on what the RA entails in terms of burden-sharing between the EU and Turkey on matters related to irregular migration, let alone the protection of migrants' rights (Kılıç 2014: 429; Elitok 2015). In a parallel vein, another major priority for the EU has been to increase the detention capacity of Turkey by funding the construction of reception centres for asylum seekers and refugees and also removal centres for illegal migrants. These are attempts to increase control over the physical mobility of migrant populations, not only from Turkey into the EU, but also within Turkey.

While the EU has been a major catalyst for substantive reform on migration and asylum policies, Turkey's EU-ization in this realm has been selective (İçduygu 2007) and subject to criticism by the EU and domestic actors. Turkey is lagging behind on its commitment towards integrated border management (EC 2013b: 64). Despite being a signatory of the 1951 Geneva Convention, Turkey currently retains the geographical limitation that restricts refugee status to asylum seekers from European countries. Reports from, the Fundamental Rights Agency (FRA), among others, have underscored that most irregular migrants apprehended in Europe arrived in Turkey legally and continued their journey into Europe in a clandestine way (EC 2013b: 65). On the one hand, Turkey has complied with the Schengen negative visa list and other requirements of the EU acquis regarding the length of stay on tourist visas by adopting more restrictive entry policies. On the other hand, Turkey remains committed to having close trade and cultural relations with non-EU countries in the region and, at times, has further extended its liberal visa policies. These points, namely the geographical limitations on who can be a refugee in Turkey and lax visa policies, imply that the entry of migrants and potential refugees is tolerated to some extent, but their access to legal status and international protection is jeopardized. Despite the EU critique and requirements, being 'tolerated but rightless', constitutes the contours of migrant illegality in Turkey and defines migrants' experiences.

Like in Morocco, international organizations have played an important role in bringing Turkey's immigration and asylum policies in line with the requirements of EU migration policies. In this context, emerging activities by UNHCR and IOM, and decisions by the European Court of Human Rights (ECtHR), have enhanced the external character of the politicization of irregular migration in Turkey. In response to the asylum influx during the Gulf Crisis, UNHCR expanded its activities in Turkey (Ozmenek 2001). Similarly, the activities of IOM in Turkey were initiated in 1991 in the aftermath of the regional crisis in the Middle East. A bilateral agreement was signed in 1995 and Turkey became a full member of IOM in 2004 in the context of a national action plan on asylum and migration. Both organizations have provided administrative support for the activities of two bureaus, namely that of Migration and Asylum and of Integrated Border Management. The UNHCR, as in the case of Morocco, has been working with implementing partners and covers their administrative costs (Ozmenek 2001). In this sense, the UNHCR is an important actor, triggering and shaping the activities of civil society in Turkey. While the UNHCR uses existing human rights activism to draw attention to asylum-related issues, the IOM has set the agenda that irregular transit migration is a problem to be managed (Hess 2012: 432). The IOM's focus on human trafficking was an important factor behind Turkey's signing of international protocols such as the Protocol to Prevent, Suppress, and Punish Trafficking in Persons, especially Women and Children, and the Protocol against the Smuggling of Migrants by Land, Sea, and Air. Furthermore, the international organization was influential in making related changes in Turkey's criminal law.

Turkey is a member of the Council of Europe. In this context, the ECtHR has been another external actor in the governance of migration. Starting with the case of Jabari v. Turkey in 2000, the ECtHR has repeatedly condemned Turkey for not respecting the principle of non-refoulement of migrants and asylum seekers.[8] The articles pertaining to detention and to non-refoulement in the Law on Foreigners and International Protection enacted in 2014, primarily aimed to be in line with the standards set by the Convention for the Protection of Human Rights and Fundamental Freedoms, commonly referred to as European Convention on Human Rights (ECHR), are designed to prevent cases against Turkey in the ECtHR. As will be detailed in Chapter 4, NGOs taking cases to the ECtHR indirectly played an important role in pushing for legal reform in Turkey (Yılmaz 2012). At the same time, their use of the ECtHR as a transnational advocacy mechanism reveals a strong external anchor in the governance of international migration in Turkey. Although not directly connected to irregular migrants' rights, ECtHR decisions have enacted the principle of non-refoulement by giving access to asylum to potential refugees who otherwise would be treated as 'illegal'.

---

8   See Mamatkulov and Askarov v. Turkey in 2005, the case of Abdolkhani and Karimnia v. Turkey in 2009 and the case of Charahili v. Turkey in 2010 (see Tolay 2012: 47; Yılmaz 2012), for exemplary decisions by the ECtHR criticizing asylum and detention practices in Turkey.

From international production of illegality to public policy

One can only grasp the production of migrant illegality in Turkey or Morocco, hence migrant experiences of illegality, by linking it to emerging forms of migration management at EU external borders. To reiterate, migrant illegality was initially a product of international dynamics. After the 1990s, there was more attention on irregular border crossings through Turkey and Morocco into Europe. This is of major concern to the EU, which pressures these countries to strengthen their border management, to establish national asylum systems in order that they be qualified as 'first countries of asylum', and to readmit third-country nationals passing through their territories into Europe. Both countries have arguably been incentivized to partly subscribe to the role of 'transit country', ironically by policing EU borders against 'secondary irregular movements into the EU.' Simultaneously, they had reasons to refrain from taking such a role.

Given their similar international contexts, the main resemblance between Turkey and Morocco is the challenge of balancing the EU's demands to stop irregular border crossings with their national interests – that is, not to become a buffer zone for immigrants – and of not worsening their relations with other countries in the region. Their relations with the EU as a sending country and a major political ally in the case of Morocco, and a former sending country and a candidate member in the case of Turkey, have influenced their cooperation with the EU on the matter of irregular migration. The main differences between the two countries have been Turkey's now fading prospects of EU membership, as well as the asylum recipient role Turkey has had to play since the 1980s, including in the ongoing Syrian conflict.

External dynamics in the production of illegality are coupled with internal dynamics of the peripheral context. The next section discusses the legal and political changes that Morocco and Turkey had to introduce in order to govern irregular migration. Note that the framework of legal changes is rudimentary and at times contested. Immigration policies in both contexts were introduced without the political will to receive immigration. Referring to Foucault, Walter Nicholls suggests: 'the enforcement of interdictions contributes to the explosion of talk, ideas, controls, and practices of illegality rather than their repression' (2013: 202). In this light, the next two sections examine how migrant illegality is produced through national laws and policies in the context of high external pressure to curtail irregular border crossings and how the issue has been subject to different forms of politicization.

2.3   Moroccan immigration politics from criminalization to integration

Since the early 2000s, trans-Saharan migration through Morocco has been represented in Morocco not only as an external dimension of the EU's migration policies, but also as a Moroccan public policy issue (Natter 2014). This politicization of irregular migration as a domestic issue is the result of the introduction of new legislation on the subject, the establishment of new institutions, and public statements of the official framing of immigration as a problem of security and criminality. Until 2013, the official discourse that Morocco is a transit country and that migrants in Morocco do not want to stay there underpinned their exclusion from the sphere of rights and membership. The extent to which a radically new immigration policy approach will replace the racialization and criminalization of irregular migration with a human rights-based integration policy is questionable. Clearly, Morocco's irregular migration policy displays evidence of rupture, at least at the discursive level, if not in practice.

Emergence of immigration policy and criminalization/

Law 02-03, the Law regarding Entry and Residence of Foreigners in the Kingdom of Morocco and Irregular Emigration and Immigration, was enacted in 2003 to improve tense relations with the EU.[9] The new law abolished earlier regulations concerning foreigners and emigrants, dating back to the protectorate period and Royal Decree in 1949, respectively (DEMIG 2015). As the name suggests, the law concerns irregular border crossings by Moroccans as well as irregular entry, stay, and exit by third-country nationals, but with little provision regarding the human rights of migrants (Belguendouz 2009: 19-20). The law regulates administrative procedures regarding the deportation of migrants and their removal to the frontal zones. Article 26 of the law prohibits the deportation of asylum seekers, refugees, pregnant women, and minors. As envisaged by the law, the Directorship for Migration and Surveillance of Borders, the unit responsible for irregular migration within Morocco, was established under the Ministry of the Interior. Coupled with EU funding for border infrastructure, Law 02-03 strengthened the mandate of the Ministry of the Interior and its securitized approach to issues concerning immigration (Wunderlich 2013: 415-6).

9   Dahir no: *1-03-196* (11.11.2003).

Considering the high number of Moroccan nationals among those crossing borders irregularly, it is surprising that there were so few discussions in the parliament regarding the substance of the law as well as a notable lack of debate among civil society actors (Feliu Martínez 2009: 351). The parliament adopted the law, together with new legislation on terrorism, in the aftermath of terrorist attacks in Casablanca in May 2003.[10] According to Belguendouz (2009: 20), civil society placed greater focus on the law on terrorism. One explanation why the law on irregular migration did not receive much criticism from opposition parties or civil society lies in the fact that irregular migration in Morocco has been publicly framed as an issue that primarily concerns trans-Saharan transit through Morocco into Europe (Natter, 2014). Even before the law, irregular migration in Morocco was presented as a sub-Saharan, rather than domestic, issue in the media (Belguendouz, 2009: 19).

The racialization of irregular migration as a sub-Saharan issue has been instrumentalized to make the law more acceptable in the public domain. Officials have justified the use of coercive measures against sub-Saharan migrants by depicting it as a fight against mafia networks controlling human trafficking through Morocco. In November 2003, the King convened a meeting on the question of migration and the surveillance of borders with the aim of combatting human trafficking.[11] As Khalid Zerouali, the Head of the Directorship for Migration and Surveillance of Borders, has explained: 'Since 2004, we have disrupted 1000 networks, it shows that we are not facing isolated cases or isolated attempts of clandestine migration but a market controlled by mafia gangs [...] Morocco is equally a leading example of a cooperation model with the North as in the examples of close cooperation we have with Spain and other countries.'[12] By mid-2005, the success of these measures was reported widely in the Moroccan media, presented in terms of decreasing the volume of clandestine migration into Spain (Valluy 2007; Feliu Martínez 2009: 350).[13]

In response to the securitization of borders, in particular in the Canary Islands, as mentioned earlier in this chapter, migrants started to engage in

---

10   Law 03-03 Regarding the Fight Against Terrorism, Dahir no: 26 (28/.5.2003).
11   See *Eriger en priorite la lutte contre les reseaux des etres humaines* ('To prioritize the fight against human trafficking networks'), *L'opinion*, 12.11.2003.
12   Interview with Khalid Zerouali, the Head of Directorship for Migration and Surveillance of Borders, Khalid Zerouali: *Le Maroc est a moins de 65% candidats a l'emigration clandestine.* ('Morocco is 65% less candidate for clandestine emigration'), *Liberation*, 15-16 July 2006.
13   See for example '37% decline in clandestine departures in the first eight months of 2005', *Liberation*, 08.09.2005.

more coordinated attempts to cross into Melilla and Ceuta. In September and October 2005, migrants were shot by Moroccan and Spanish border guards during attempts at the borders between Morocco and the Spanish enclaves, proving the human cost of these coercive measures (Belguendouz 2009: 21; Migreurop 2006). Moroccan security forces unlawfully removed large groups of undocumented migrants to the no man's land between Algeria and Morocco before and after the clashes (GADEM 2007: 16). The Ceuta and Melilla scandal led to increasing international attention to the treatment of international migrants on Moroccan soil. The coercive practices violating national and international laws have become much more visible and have been criticized by domestic and international actors. The event not only showed the human cost of border controls in the absence of fundamental rights, but it also paved the way for contestations from within and outside Morocco. In other words, criminalization from above gave rise to an emerging 'politicization from below'.

2005 was a turning point for the expansion of civil society activities concerning irregular migration (Semeraro 2011: 55; Jacobs 2012). In general, the increase in civil society activism has been part of the political liberalization of Morocco since the 1990s (Cavatorta 2009). Note that it was a period when more funding opportunities, especially from the EU, were available for NGOs working on irregular migration issues in Morocco (Dimitrovova 2010). Awareness of and sensibility towards the vulnerable situation of irregular migrants passing through Morocco into Europe started before 2005. International organizations such as the IOM and the UNHCR, but also humanitarian organizations such as Doctors Without Borders (MSF) and Caritas had already initiated their activities on irregular migrants in Morocco. MSF Spain started to operate in 2003 in Tangier, in 2005 in Nador and a couple of years ago in Casablanca and Rabat (author interview with MSF, Rabat, April 2012). Caritas has been operating with vulnerable populations in Morocco since the 1950s, and their activities included irregular migrants as this group became more visible in urban centres such as Tangier, Rabat, and Casablanca since 2002. Their reception centre in Rabat was opened in 2005, as more migrants were coming to Rabat as a result of intense deportation practices (author interview with Caritas, Rabat, July 2012). Civil society working on immigration-related issues was rather nascent in 2005 (Feliu Martínez 2009: 352) but proliferated in a dynamic fashion.[14] In a context of widespread rights violations, civil society actors have become increasingly vocal with regard to migrant rights.

---

14   The structure and main activities of these CSOs and their relations to the state are further discussed in Section 3.4.

Even after the events at Ceuta and Melilla, authorities have been unwilling to hear civil society's and migrants' demands for the recognition of the rights of undocumented migrants under national and international law. The use of the label 'transit country' justified the security-oriented legal framework, but also the practices on the ground that are particularly restricting with regard to irregular migrants' access to fundamental rights such as non-refoulement, access to asylum, access to healthcare, and minors' access to education. A related justification for the lack of inclusionary and integration policies for migrants is the low capacity of the Moroccan state to receive migrants. Most officials interviewed during my fieldwork in summer 2012 underscored that Morocco is not a country of immigration in terms of economic development: 'There is nothing for migrants here but it is seen as better than Gabon [...] What can Morocco offer to migrants? Best scenario is exploitation' (author interview, Rabat, June 2012). The widespread conviction is that Morocco is a victim of its geographical position and has no option other than to follow European policies. A member of the Moroccan parliament from the ruling Justice and Development Party underscored this: 'We feel like we are in the right direction. Because of its geographical position, Morocco must implement European laws. Morocco does not have the means [...]' (author interview, Rabat, September 2012). While the need for regional cooperation rightly prevails in the discourses of state and non-state actors, putting the responsibility on European actors and presenting them as the source of the problem also becomes a strategy to deny migrants' rights. As articulated by an official: 'There is violence, extreme poverty, but it is not up to Morocco to find a solution. We need a global, regional, international strategy' (author interview, Oujda, September 2012). The use of a 'transit card' (Hess 2012: 436) not only works to increase Morocco's leverage towards the EU, but also emphasizes its role as a country of transit rather than immigration, lacking the necessary capacity to deal with migratory flows.

The humanitarian and advocacy activities by civil society, intensified in the post-2005 period, have been at odds with the official state perspective. Civil society organizations (CSOs) have become part of the governance of irregular migration in Morocco by undertaking the integration task that the state is explicitly unwilling to perform (Natter 2014). Hence, they have contested the production of migrant illegality and have worked towards the decriminalization of irregular migration. Civil society has not only created channels, albeit limited, for migrants' de facto access to rights and services, but it has also provided a political sphere for irregular migrants to claim rights as members of society. Plus, migrants removed from the

Spanish frontier have become much more visible in urban settings, and have also started to organize among themselves. As explained in Chapter 3, civil society activities and migrant activism have contributed significantly towards what I call a rupture in Moroccan immigration policy.

The emergence of civil society working on irregular migration issues has to be contextualized in the wider political and institutional liberalization process. This liberalization process of associative life has been extended since 1999, under the reign of Mohammed VI (Sater 2007: 160-161). Introduced in the aftermath of the Arab revolts, the 2011 Constitution included articles such as Articles 12 and 30 on the human rights of foreigners. Article 161 of the Constitution reformed and enabled more independent grounds for action by the National Council of Human Rights (CNDH) (Cherti and Grant 2013: 5-6). Despite its fragile and highly criticized position, the CNDH has played a key role in shaping what is called a 'radically new immigration policy'.

Towards integration?

In the context of growing national and international critiques on the treatment of irregular migrants in Morocco, the report on the human rights of foreigners in Morocco, presented by the CNDH to the King in September 2013, heralded a clear turn in the country's migration policies. Acknowledging that Morocco has become a land of immigration, the CNDH recommended a set of policies to facilitate legal and socio-economic integration of both asylum seekers and migrants (see CNDH 2013). As mentioned, the mobility partnership agreement signed with the EU in June 2013 recommended the introduction of a new asylum and international protection system in Morocco. These critiques and recommendations led to a paradigmatic change in Moroccan immigration policies, initiated by King Mohammed VI, who underscored, in his royal discourse, that Morocco is becoming a land of immigration for sub-Saharans and Europeans alike and that there is need for a new policy perspective.[15]

Following the King's initiative, in November 2013, the government announced a regularization campaign targeting immigrants with irregular legal status in Morocco. This regularization programme lasted throughout 2014. The years of residence required for eligibility alternated between different categories of migrants:

---

15   See Royal discourse on the occasion of 38th Anniversary of the Green March, 06.11.2013, Retrieved 15.05.2015, from http://www.map.co.ma/fr/discours-messages-sm-le-roi/ sm-le-roi-adresse-un-discours-la-nation-l%E2%80%99occasion-du-38eme-anniversaire

The exceptional operation of regularization concerns foreigners with spouses from Moroccan nationality living together for at least two years, foreigners with foreign spouses in legal status in Morocco and living together for at least four years, children from the two previous cases, foreigners with employment contracts effective for at least two years, foreigners justifying five years of continuous residence in Morocco, and foreigners with serious illnesses who had arrived the country before December 31 2013.[16]

As advertised by policymakers, this practice has made Morocco the first among developing countries engaged in regularization campaigns. Officials emphasize that the new policy envisages a humanitarian approach to asylum and immigration that respects international norms and the human rights of migrants. Clearly, this indicates a shift from previous official discourse that Morocco lacks the resources to deal with immigrants, who are allegedly stuck on Moroccan soil on their way to Europe. For instance, Mustapha Kassou, a member of the CNDH publicly stated: 'This is a sinuous but irreversible path. Our country has the means to achieve socio-economic integration of migrants present in its territory' (cited in Lemaizi 2013).

The launch of the new policy was followed by institutional and legal changes and a possible rapprochement between authorities, international organizations, and civil society in Morocco. The Ministry Responsible of Moroccans Abroad was renamed the Ministry Responsible for Moroccans Abroad and Migration Affairs.[17] This decision was welcomed by NGOs, as it meant that a ministry in charge of social affairs, rather than security issues, would now take charge of the regularization campaign (Alioua 2013). The department has expressed an intention to collaborate more closely with civil society organizations active in the field of human rights. The UNHCR and IOM both confirmed increasing coordination with the new ministry in follow up interviews in May 2014. Foreigners' Offices have been created to operate the regularization programme. An ad hoc commission was formed to work on the national asylum law. An asylum bureau opened in Rabat to coordinate with the UNHCR in processing asylum cases.

16 Retrieved 15.05.2015, from http://www.marocainsdumonde.gov.ma/actions-du-minist%C3%A8re/le-maroc-lance-du-1er-janvier-au-31-d%C3%A9cembre-2014-une-op%C3%A9ration-exceptionnelle-de-r%C3%A9gularisation-des-%C3%A9trangers-en-situation-irr%C3%A9guli%C3%A8re.aspx
17   Referred to as the Ministry of Migration Affairs, hereafter.

The introduction of a new approach to migration and the regularization campaign has developed rapidly, reflecting the decisive role of the
King in Moroccan politics (Cavatorta 2010: 17). It should be noted that
this reformist turn in Morocco's immigration policy was unexpected. As I
started my fieldwork in Morocco in April 2012, the demand for regularization was rather implicit, and stakeholders interviewed were pessimistic
about a positive change in Morocco's immigration policy. Regarding the
question of demands for regularization, the response was very clear that
migrants in Morocco are in transit and that Morocco cannot be 'a solution
for exchange' for those migrants who, in the first place, wanted to reach
Europe and who stay in Morocco only because they cannot reach their
end destination. The striking turn in the tone of officials as well as policy
after September 2013, however, was clear. Public speeches by the Minister
of Migration Affairs, Anis Birou, underscored the radical change in the
official discourse. For example, during an international meeting on the
new policy, he stated that:

> Morocco, because of this new policy, will save thousands of lives. We all
> want to prevent that there are going to be new Lampedusas. We all want
> that this new immigration policy announced in Morocco will go beyond
> the borders of Morocco. This new migration policy of Morocco does not
> only concern Morocco… We believe that this is a shared responsibility, we
> are all assuming this responsibility in giving migrants a second chance
> to realize their dreams, instead of the hell of crossing the Mediterranean,
> to realize the Moroccan dream.[18]

While the impact of the new migratory approach in terms of remedying
migrants' experiences of exclusion has yet to be seen, most analyses lo-
cate this recent turn in Morocco's migration policy within the country's
geopolitical strategy for forging firmer relations with the EU, and with
African countries in order to compensate for its absence from the African
Union. It is also acknowledged that this process is linked to improvements
in fundamental rights as envisioned in the 2011 Constitution, in a period
when Morocco is acknowledged as a country of immigration rather than
merely a transit zone. Despite the top-down character of the new migratory
policy initiative, this policy initiative should be seen as a response to the

---

18   Author's notes from the meeting "The new migration policy in Morocco, which strategy of
integration?" organized by the Ministry in Charge of Moroccans Abroad and Migration Affairs,
IOM, Confederation of Switzerland, 11-12 March 2014, Rabat, Morocco.

ongoing international and domestic criticism towards the Moroccan state for denying the rights of irregular migrants.

High criteria for eligibility and the uncertainty awaiting those who are not regularized are the most criticized aspects of the regularization campaign. As a partial response, authorities loosened criteria for regularization in order to include women, minors, and Syrian refugees as vulnerable groups, as well as activists and leaders of informal migrant associations. By the end of 2014, 'close to 17,918 one-year residence permits were granted' (almost half of them to Senegalese and Syrians, followed by Nigerians and Ivoirians) (Martin 2015). Questions on the implementation of the regularization campaign cast doubt on whether the new policy will ensure human rights and the integration of migrants or if it will lead to another form of control over migrants, for example by collecting personal information from migrants, including those who are not eligible for regularization. NGOs are equally concerned about how the personal information they provide will be used and if it will be shared with authorities such as the EU for readmission or other purposes. Furthermore, the collaboration with NGOs might create bias against those groups not involved in NGO activities (Chaudier 2013). There has been scepticism about the new policy approach's effectiveness to end coercive measures (Chaudier, 2013). While there has been a rupture in the criminalization of migrants' presence on Moroccan territory, securitized measures continued, crystallized through removal practices that led to severe injuries and deaths, especially along the border, throughout 2014 and at the conclusion of the regularization programme (Belghazi 2015).

To summarize, since the early 2000s, irregular migration in the Moroccan context has been conceptualized as criminal activity. The official stance that Morocco is a transit rather than a migrant receiving country and a victim of its geopolitical position continued until late 2013. Since September 2013, there has been a discursive turn in Moroccan immigration policies. Highlighting Morocco as a case of rupture, this section has clarified that the policy background is characterized by the criminalization of irregular migration as well as the gradual acknowledgement of irregular migrants' right to stay. Chapter 3 will go further to explain how this particular criminalization and politicization is interlinked with experiences of illegality in terms of exclusion, but also gives rise to particular forms of informal incorporation through migrant mobilization. It also makes the case that bottom-up politicization unfolded, directly or indirectly, as a result of the rupture in immigration politics.

2.4   Migrant illegality as Europeanization in Turkey

Turkey's transition from the absolute absence of any immigration policy to the adoption of an immigration policy as part of the EU accession process has been gradual. The institutionalization of migration governance initially emerged as a response to incoming asylum flows and evolved as a matter of Europeanization. Parallel with the adoption of techniques to govern the EU's external borders, the post-2000 period has witnessed a transition in scattered immigration policies in Turkey. Concurrently, immigration legislation and institutionalization in Turkey have mainly been discussed in public and policy circles, within the technicalities of the EU accession process. This section re-evaluates what is documented in the literature as the Europeanization of migration and asylum policy (İçduygu 2007; Özgür and Özer 2010; Ozcurumez and Şenses 2011) as a case of the institutionalization of migrant illegality.

Emerged as refugee, developed as an EU issue

Until the mid-1990s, the Turkish state was not actively involved in regulating immigration flows. In contrast with Morocco, the issue of asylum was the initial object of governance in Turkey, rather than the problems of irregular or clandestine migration. For example, in Hess' study conducted in the early 2000s, potential informants could only relate to the research theme 'transit migration' when researchers mentioned the name 'refugees' (2012: 431). The first national legislation on asylum appeared in 1994. Fearing mass inflows during the Gulf Crisis and at the peak of the Kurdish armed conflict in the eastern part of the country, authorities introduced a regulation on refugee status determination. According to the Regulation on the Procedures and the Principles Related to Mass Influx and the Foreigners Arriving in Turkey either as Individuals or in Groups Wishing to Seek Asylum either from Turkey or Requesting Residence Permits with the Intention of Seeking Asylum from a Third Country,[19] the Ministry of the Interior became the final decision-making body for refugee status determination in collaboration with the UNHCR. While the 1994 Regulation marks a transition into international norms (İçduygu and Bayraktar Aksel 2012: 40), the post-1994 period is also characterized by rights violations by Turkey, especially the right to non-refoulement, and by increasing cases against Turkey at the ECtHR (Kirişci 2012: 67-8). The 1994 Regulation introduced administrative

19   Referred to as the 1994 Regulation, hereafter.

procedures requiring applicants to register with the police within five days of arrival and to reside in cities designated by the police. The 2006 Implementation Directive removed this temporal clause and replaced it with a 'reasonable time period'. Still, officials tended to strictly implement these measures and increasingly deported potential refugees failing to meet rigorous administrative requirements (Kirişci, 2012: 67). In other words, there were, arguably, few differences in the treatment of potential asylum seekers and those seen as 'illegal' before the law.

After the initial phase leading towards the adoption of international norms on asylum, the signing of the Accession Partnership Agreement with the EU in 2001 pushed for legislative and institutional changes in the field of asylum and migration in Turkey. The National Security Council issued a resolution on irregular migration in 2002, and the Strategy Paper for the Protection of External Borders in Turkey was adopted in 2003 (DEMIG 2015). More restrictive visa policies have been adopted in line with the Schengen negative visa list (DEMIG, 2015). Legal activism in the context of EU-led reforms targeted what might be called Turkish immigration policy, which was regulated through various legislation such as the Passport Law, the Law on Residence and Travel of Foreigners in Turkey, and the Citizenship Law. The adoption of the Law on Work Permits of Foreigners, changes in the law regulating the acquisition of citizenship through marriage, and harsher sentences introduced for human trafficking and smuggling in 2003 were among the important and unprecedented legal changes in the field of international migration in Turkey in the post-2000 period. Some of these legal measures were envisaged within the adoption of the EU acquis, and others were reactive measures to the changing mobility dynamics in Turkey. For instance, the introduction of a three-year waiting period for the acquisition of citizenship through marriage was a response to the perception that female migrant workers with post-Soviet origins were legalizing their stay through marriages of convenience (Bloch 2011: 508).

The period between 2003 and 2008 is characterized by legal activism in the context of Europeanization as well as increasing civil society awareness. At the implementation level, enforcers were granted considerable space for discretionary power. The wide interpretation of notions such as Turkish traditions, political requirements, and violating peace and security as grounds for detention and deportation led to various forms of human rights violations (Dardağan Kibar 2013; Yılmaz 2014). The case of Festus Okey, a Nigerian asylum seeker killed by a police gun while being detained in the Beyoğlu Police Station in Istanbul in 2007, has been a highly visible example of such (rights) violations. The continuation of rights abuses during the trial of this case triggered further civil society activism and led to rising awareness among academics, lawyers, and other civil society actors about the question of asylum and immigration in Turkey.

As is further detailed in Chapter 4, existing as well as newly established human rights organizations and other civil society actors developed an interest in immigration and asylum issues in the post-2005 period. As commonly agreed by several NGOs interviewed, the EU accession process has provided the basis for the emergence of civil society organizations working on immigration issues in Turkey in terms of opening a political space for – and making available – funding opportunities. Humanitarian and advocacy organizations have become more involved by becoming service providers for the UNHCR and thus mainly served asylum seekers. In this sense, the move from no policy to the adoption of a policy in the post-2003 period has made irregular migration a subject of governance with the involvement of multiple external, state, and non-state actors. Features of this emerging governance included a scattered legislative framework, a security-dominated approach to irregular migration, a lack of public awareness or debate on the subject, rights violations in implementation, and increasing civil society critiques. The extent to which Turkey's new asylum and migration legislation, and the institutionalization around it, can remedy rights violations and change forms of politicization around immigration is yet to be seen.

New legislation and the institutionalization of migrant illegality

The main motivations for the institutionalization of immigration and asylum governance in the post-2008 period were the commitment to the adoption of the EU acquis, preventing the ECtHR's decisions against Turkey, and growing international and domestic civil society activism leading to critical reports on rights violations (Kirişci 2012: 77; see e.g. HCA 2007). In close cooperation with particular EU states, the UNHCR, and the IOM, the Migration and Asylum Bureau and the Bureau for Border Management were established in October 2008 under the Ministry of the Interior. The establishment of these two bureaus is indicative of the institutionalization of migration bureaucracy in Turkey as well as the first steps of the politicization of immigration issues. The main mission of the Migration and Asylum Bureau was to draft the Law on Foreigners and International Protection (LFIP). Prepared in regular consultation with stakeholders such as CSOs and academics, the draft law was made public in 2011, and the LFIP came into force in April 2014, a year after its enactment. Legal and institutional changes envisaged by the law arguably heralded a new phase in the governance of immigration and asylum in Turkey. The process has led to the institutionalization and emergence of a bureaucratic cadre focused on immigration in the post-2008 period.

As the name suggests, the LFIP includes foreigners' law and asylum law. It brings together formerly scattered pieces of legislation on entry, stay, and the deportation of foreigners. For the first time, Turkey's asylum policy is codified as law, as opposed to secondary legislation, which, in previous periods, mainly referred to regulations. As a major institutional novelty, the law centralizes the policymaking and implementation in the field of international migration, and asylum under the Directorate General of Migration Management (DGMM). Before the LFIP, various state bodies were simultaneously responsible for policies concerning immigration. The most prominent of these were the Department of Foreigners, Border and Asylum under Directorate of General Security of Ministry of Interior, and the Deputy Directorate General for Migration, Asylum and Visa under the Ministry of Foreign Affairs. As envisioned by the law, DGMM and its subsidiary organizations, which are institutionalized at the provincial level, will gradually take over responsibilities from the Turkish National Police (TNP). As in the case of Morocco, the EU's support institutionally strengthened the Ministry of the Interior, but the organization has institutionalized under a civil bureaucracy rather than the police department and the military. The development of the civil bureaucracy arguably led to the strengthening of human rights-based approaches in immigration policymaking, along with the security agenda that dominated irregular migration discussions since the early 2000s.

The law-making process has revealed a gradual change in relations between rights-based NGOs in the field and state institutions. In the process of law-making, the state recognized the presence and importance of non-state actors and their experience in the governance of migration in Turkey. Despite tense relations between civil society and the state due to reports criticizing deportation and detention practices, the opening of dialogue with civil society has been at the core of the law-making process (author interview with HCA, Istanbul, November 2013 and with Amnesty International Turkey, Ankara, November 2012). The Migration and Asylum Bureau has welcomed the exchange of ideas regarding the content of the law. During the process of legislation in the parliament, the presence of NGOs in commission meetings was an important aspect of law-making. The interviewee from Amnesty International articulated the following:

They are the ones organizing meetings. We received invitations from them. We do not receive many invitations from state institutions as CSOs. As state tradition, we do not have a participatory state tradition in any subject. Same goes for migration. This happened because of the vision and individual sensibilities of bureaucrats in the Bureau. Also, the Minister of Interior at the time was more open to dialogue with civil society. This also encouraged the bureaucrats. As a result, we were invited to several workshops and consultations.

The same interviewee also added that civil society's presence in Parliamentary Commission meetings was not by invitation, but was due to their insistence on participating: 'I called the Commission to ask if we could participate at the meeting. They first said no, they said, "You need approval of the head of the commission". We had to act quickly; in the end, we forced them to invite us. We could receive the written permission.' However, limitations on NGOs' participation in certain meetings and short consultation periods indicate a top-down inclusion process. In this sense, it differed from Morocco, where civil society and migrant organizations had to carve out their political space.

Along with procedural changes, there has been a change in terms of the framing of the issue of irregular migration in particular and of immigration in general in the post-2008 period. Arguments pertaining to incapacity to deal with migration also hold in Turkey and were applied in a similar way to the case of Morocco. Officials have maintained their concern over burden sharing with the EU and security aspects of migration (largely discussed elsewhere, see Kirişci 2012; Tolay 2012: 54; İçduygu and Üstübici 2014: 54-5). As articulated by one official from the police department: 'If we agree on readmissions, our streets will be full of foreigners; we cannot walk around comfortably.' At the same time, the perspective has shifted from a securitized to a human rights approach. Atilla Toros, a well-known bureaucrat in the field of migration and asylum, and the first head of the DGMM, publicly stated that he had visited detention centres and had spoken with asylum seekers in satellite cities during the preparation of the LFIP. By saying, 'We looked in the eyes of asylum seekers while writing these laws,' he alludes to the degree of shift from a purely state-centric to a more human rights-based perspective. The increasing number of reports by the Human Rights Commission in the Turkish Parliament also exemplifies the growing interests to protect the rights of migrants and asylum seekers, even before the large number of Syrians settling in Turkey (see reports by Turkish Parliament Human Rights Inquiry Committee 2010, 2012, 2014).

Another motivation for the law was the economic aspect of immigration. The overall rationale of the law published by the Ministry of the Interior underlines 'Turkey's climbing economic power' as an attraction for migratory movements.[20] In a parallel vein, officials interviewed widely referred to Turkey's 'own dynamics', alluding to the conviction that Turkey required these reforms regardless of EU accession. The term 'own dynamics' refers to the growth of the Turkish economy since the economic crisis in 2001. As the macroeconomic variables indicate, Turkey is much more integrated into the global economy than Morocco. Thus, Turkey's immigration experience is related to the country opening up to the global economy and its broad informal sector (Toksöz, Erdoğdu and Kaşka 2012; İçduygu and Yükseker 2012). The informal sector has grown over many decades and has absorbed low-skilled workers from different parts of the country who are excluded from the formal sector.

According to the Ministry of Labour and Social Security, the informal economy constitutes nearly half of the total economy. The Turkish Statistical Institute estimates that unregistered informal employment comprises 40 per cent of the total employment (Arca 2013). Within this picture, the unregistered foreign labour force has predominantly been informally employed in small and medium-sized workplaces in construction and related industries, as well as in the leather and textiles industries, agriculture, care, and tourism. (Toksöz, Erdoğdu and Kaşka 2012: 72-6).

On the one hand, the relaxation in visa policies since 2010 shows Turkey's ongoing 'laissez-faire' approach to irregular labour migration and indirectly to transit migration into the EU. On the other hand, recent changes in visa policies are aimed at curtailing circular mobility and at registering those overstaying in Turkey. In line with the EU acquis, the law requires that 'the duration of stay provided by the visa or visa exemption shall not exceed 90 days within 180 successive days.' The EU accession process and the increasing visibility of irregular labour migrants in certain sectors have been the main motivations for changing the visa policies. This legal change was followed by an exceptional regularization scheme implemented in the summer of 2012 to give a chance to those who entered the country before the illegalization of multiple entries. It was a one-time amnesty whereby migrants with a legal entry could pay fines for the time they overstayed and apply for a six months exceptional, non-renewable residence permit. Note that it only applies to those overstaying their visa in exchange for rather high fees. The initiative was designed to remedy the change in the

20   See "Overall Rationale", *DGMM*, Retrieved 15.05.2015 from http://www.goc.gov.tr/icerik3/overall-rationale_913_975_977

entry laws (see Article 11 of the LFIP), rather than to forge a regularization campaign as we have seen in the case of Morocco.

The rationale behind the law acknowledges irregular migrants' economic presence in the country. The Turkish state indirectly admits that there have been violations, particularly in the context of the deportation of irregular migrants. Meanwhile, it is not possible to talk about either an official demand or a conviction for the need of a foreign labour force.[21] The only sector where the need for migrant labour has been acknowledged by officials interviewed at various levels has been child and elderly care. At the same time, the content of the law does not radically extend the rights of irregular migrants. The law's main impact on the lived experiences of illegality can only truly be understood in practice. However, as a written document, the law aims at providing a clear filter between asylum seekers, legal migrants, and the illegal (Tolay 2012). An official from the police department explained these distinctions and the aim of the law as follows: 'The food comes into the body, if it is good (legal) it is digested, if it is bad (illegal), it is thrown away.' As this metaphor suggests, the content of the law arguably aims to reinforce the distinction between asylum seekers' legitimate right to stay and the illegitimate presence of irregular migrants. The LFIP does not lift the geographical limitation on who can be admitted as a refugee in Turkey. However, provisions in the law ensure the principle of non-refoulement, access to asylum, and enjoyment of fundamental rights by asylum seekers and refugees. The LFIP contains no similar provisions regarding irregular migrants' access to healthcare and education.

The issues of detention and deportation are the most criticized and hence politicized aspects of irregular migration management in Turkey. Reports have focused on the widespread use of detention, long detention periods, conditions of detention centres, and unlawful deportations and detentions because of the problems inherent in the functionality of the international protection system (HRW 2008, 2007; SRHRM 2013; for an extensive list of report on the subject, see Grange and Flynn 2014: 19). As emphasized by the UN Special Rapporteur on Human Rights of Migrants, detention appears to be a migration control technique rather than a measure of last resort (SRHRM 2013: 10-11).

In response to critiques from different actors, the law clearly aims to standardize the treatment of foreigners by leaving less room for discretion in

---

21   There has been an ongoing debate on facilitating work permits of foreigners in certain sectors as well as in the case of Syrians. By January 2016, these discussions did not turn into a concrete policy.

the hands of authorities, especially with respect to deportation and detention decisions (Dardağan Kibar 2013). When compared to previous legislation, Articles 54 and 55 of the law provide more grounds for justifying deportations in cases of irregular entry, stay, and work. At the same time, it provides protective measures to certain groups in vulnerable situations. The legal basis for detention is provided for the first time, and the terms of detentions are clarified. In direct response to ECtHR decisions against Turkey, the law ensures procedural guarantees, and the right to appeal to decisions, entry bans, detentions, and deportations. In other words, migrants and/or their legal representatives are given time to leave the country and the possibility to go to administrative courts to contest authorities' detention and deportation decisions. However, there are exceptions in the law that designate conditions under which the legal period to leave Turkey may not be granted. These exceptions include abstract clauses such as posing 'a public order, public security, public health threat' and give authorities a degree of discretionary power and the capacity to legitimize immediate deportations; thus, they can potentially preclude irregular migrants' access to procedural guarantees and jeopardize their right to stay in the country. As a result, while the law brings important novelties, especially procedural guarantees, regarding irregular migrants' right to stay, certain clauses on discretionary power may lead to the continuation of arbitrary practices that violate human rights.

The LFIP also brought unprecedented novelties such as permanent residence permits or articles mentioning the integration of foreigners and asylum seekers. Notably, there were few political debates and hardly any negative views on this emerging immigration policy realm during the preparation and legislation processes. This resonates with the general lack of public discussion and parliamentary discussions on the subject of irregular migration and asylum (Tolay 2012). Interviews with HCA confirm that despite the increasing awareness that Turkey is becoming a country of immigration, immigration has not yet become a political or electoral issue that concerns the general public or their opinions (author interview, Istanbul, November 2013). As of the end of 2014, prior to the Syrian conflict, the issue had not become part of high politics, in the sense that political parties would have differing stances on the question of immigration.

In this context of lower levels of politicization, media coverage tended to reproduce stigmas around certain migrant communities, rather than inform public opinion on socio-political and human rights aspects of the issue. Informants underscored that media attention to the subject has been limited to accidents and casualties along the land and sea borders. The sparse media attention on immigration is likely to change with the Syrian

crisis. Even in the case of Syrian refugees, the initial media attention has
been limited considering the incredible number of Syrian refugees (Düvell
2013). Meanwhile, the Syrian conflict has gradually altered the low political
profile and external character in relation to asylum and migration issues in
Turkey. In the several provinces where Syrian refugees are most visible, there
has been evidence of discontent regarding their influx (Şimşek 2015: 59-60).

The LFIP has arguably re-defined migrant illegality in legal terms and
introduced procedures and rights that are more lenient with asylum seekers
while being tougher on irregular migrants (Tolay 2012: 52). The outcome of
the legal changes in terms of redressing heavily criticized human rights
violations can only be seen in their implementation. The rationale of the
law recognizes the presence of irregular migrants in the economy and shifts
away from a security approach to one that is concerned with international
mobility in general. However, the content of the law provides no rights for
irregular migrants, aside from procedural guarantees in cases of detention
and deportation. The law-making process has clearly opened up a dialogue
between state actors and civil society. However, the new legislation and
institutions, that is to say, the shift from no policy to policy on immigration
and asylum, did not necessarily alter the low levels of politicization around
the issue. This trend of depoliticization has changed with the arrival and
increasing visibility of Syrian refugees in Turkey, especially after 2014.

To conclude the section, the discussions and practices around irregular
migration in Turkey are incorporated into asylum and Europeanization
discussions. Scholarly research has framed the policy transformation as a
case of Europeanization. This section has argued that what is disguised as
Europeanization has, in fact, been the institutionalization of migrant illegality. The section has explained the rather informal character of immigration
policy and the depoliticized nature of migrant illegality in Turkey. I have
suggested that relatively lower degrees of politicization have characterized
the governance of irregular migration. Chapter 4 will further explore the
impact of relatively low levels of politicization of irregular migration on
migrant incorporation.

Conclusion: From international production of migrant illegality to migrant incorporation

Focusing on the policy and institutional levels, this chapter has sketched
the diversity of actors and contextual factors contributing to the production of migrant illegality in two countries. Both Turkey and Morocco have

intrinsically participated in the EU migration regime. In both contexts, irregular migration was initially an aspect of their changing emigration flows to the EU and later became a policy concern regarding incoming flows. The volume, source countries, and profile of incoming migrants differ from one context to another. What is comparable, as I have suggested, is the emergence of irregular migration as a subject of governance in Turkey and Morocco, through similar techniques of producing migrant illegality as well as the countries' comparable positions within the international context.

It is undeniable that the EU has played a major role in rendering irregular migration a subject of governance in its periphery. The notion of a 'transit country' is important for understanding the impact of the international context on the production of migrant illegality in peripheral contexts. The countries identified as transit have taken measures to control mobility along their borders with the EU. Ironically, these countries are labelled as 'transit' due to measures they have introduced in collaboration with the EU to 'stop transit'. In peripheral contexts such as Turkey and Morocco, migrant illegality was initially a by-product of the political will to stop irregular entries into the EU. This has led to the increasing involvement of the EU in the border infrastructure of the transit countries as well as the increasing activities of international/intergovernmental organizations such as the UNHCR and the IOM. Additionally, it has produced changes in the legal infrastructure of transit countries.

This preoccupation with securing EU borders has had diverse outcomes. As is widely shown in the literature, rather than eradicating irregular border crossings, these measures resulted in costlier and riskier transit movement and caused migrants to spend more time in transit countries. As a result of this process, authorities instrumentalized the label 'transit country' to suspend the human rights of migrants that are allegedly on their way to Europe. The construction of certain countries as transit contributed to state discourses that sidelined their responsibilities towards irregular migrants (Oelgemöller 2011: 415). This resulted in the growth of a foreign population with no legal status, hence with no rights, in transit zones.

This process resulted in the introduction of restrictive policies, not only at border zones, but also apparent in internal migration controls. Sections 2.3 and 2.4 revealed that both Turkey and Morocco introduced restrictive legal measures to control irregular migration. Irregular migration as a policy issue was arguably more problematized and criminalized in Morocco. Conversely, in Turkey, labour aspects of irregular migration went hand in hand with security aspects. Restrictive policies and harsh enforcements have led to human rights violations and, consequently, to international and domestic critiques in both contexts. After years of denying responsibility for the rights of irregular migrants on its soil, in 2013, Morocco shifted its policies to recognize irregular migrants' right to stay and integrate. In parallel, Turkey introduced its first comprehensive law on asylum and foreigners in the same year. Since then, immigration policies in Turkey have gone through a process of gradual transition. EU-led reforms and state efforts have aimed at striking a balance between the ongoing, albeit slow, EU accession process and the increasing numbers of incoming refugees and migrants until 2014. The dynamics of asylum and migration policy-making in Turkey has changed since then, with fading EU membership prospects and Turkey's becoming the country receiving the largest number of refugees in the world. Nonetheless, the impact of the EU has not faded away completely.

Given the similar emergence of the issue of irregular migration in the political agenda despite different levels of politicization of the issue, Morocco and Turkey provide suitable comparative cases to explore the impact of the interrelation between external and domestic factors on migrant illegality. Building on the conclusions of Chapter 2, Chapter 3 on Morocco and Chapter 4 on Turkey question how the exclusionary practices vis-à-vis migrants have impacted migrants' experiences of incorporation at the levels of policy, discourse, and practice: What roles does enforcement by the bureaucracy, market, and civil society play in the incorporation of migrants and the definition of available strategies for migrants' access to rights and legal status?

3 Morocco as a case of political incorporation

Introduction

Chapter 2 characterized the governance of irregular migration in the Moroccan context in terms of the external pressure to secure European borders, the absence of political will, and a clear market demand for immigration since Morocco is still a country of emigration. Despite exclusionary discourse with respect to irregular migrants, there is a 'radically' new immigration policy initiative, albeit very recent. The regularization of migrants without legal status in Morocco has been a major aspect of the new immigration policy. While the outcomes of the new immigration policy initiative are yet to be seen, migrants' testimonies reveal the gradual but drastic change in the visibility of migrants in the social and political spheres. As observed by a member of a migrant community for sub-Saharan migrants, 'it was

Figure 3.1   A protest by migrants in the streets of Rabat, 'Halt Raids, we are in Morocco, we live in Morocco → we love Morocco'

Source: Unknown. The picture has been used on several occasions since 2012. See for instance, Le Gadem devoile la liste des lieux de detention des migrants au Maroc [Gadem disclose the list of detention places of migrants in Morocco]. Retrieved 15.03.2015 from http://www.medias24.com/SOCIETE/152908-Le-Gadem-devoile-la-liste-des-lieux-de-detention-des-migrants-au-Maroc.html#sthash.cM7mFIU5.gbpl

impossible to walk in the street back in 2005,' in the aftermath of events in Ceuta and Melilla. During a meeting in March 2014, 'The new migration policy in Morocco, which strategy of integration?' organized by the Ministry in Charge of Moroccans Abroad and Migration Affairs, a sub-Saharan migrant in the audience addressed the Minister directly, saying that he applauds the fact that children of irregular migrants are currently being admitted to primary schools, but that the curriculum is not suitable for Christian pupils. How can we account for this change? That is, how do undocumented migrants raise their voices as political actors, given the official discourse and legal framework that have, until now, criminalized their presence on Moroccan soil?

The chapter discusses policies and practices that have pushed migrants to exclusion and further marginalization and others that have enabled their social and political incorporation. Earlier research and reports have mostly focused on migrants' living conditions during their journeys to the EU and their access to fundamental rights (AMERM 2008; Cherti and Grant 2013; Alioua 2008; Pian 2009). The chapter explains how migrants of irregular status experience legal, economic and social exclusion, and negotiate their rights (to stay in the territory) through mobilization practices aimed at acquiring rights and access to legal status.

The first section explains migrants' experiences of exclusionary prac- tices of deportation, which have given rise to growing criticism, especially since 2005. The second section shows mechanisms through which migrant illegality is reproduced, resulting in exclusionary practices at different stages of the migration experience, such as settlement and labour force participation. Here, I question the connection between migrant illegality, formal exclusion from the body of membership, and informal inclusion in the labour market, widely referred to in the literature (Calavita 2005; Garcés-Mascareñas 2012). The findings highlight that exclusion is never absolute and is always negotiated on the ground. Regarding possibilities for bureaucratic incorporation, access to both healthcare and education are scrutinized to reveal mechanisms of bureaucratic incorporation and to highlight the role of civil society mobilization in enabling the access to certain fundamental rights as well as migrants' visibility in the social and political spheres. The role played by civil society, including international, Moroccan, and migrants' associations, is extensively discussed with respect to the question of mobilization for the rights of irregular migrants. The last section looks closely at how access to rights and legal status is negotiated through mobilization for the rights of irregular migrants and how irregular migrants themselves have become vital to this civil societal network. The emergence of a vibrant civil society in Morocco and the alliances built between Moroccan and migrants' associations increased migrants' visibility as rights-bearing subjects seeking membership on Moroccan soil. The chapter argues that immigrants of irregular status in Morocco have been incorporated as rights-seeking political actors despite the physical, economic, and social exclusion they have experienced.

3.1 Deportability as part of daily experience

The literature extensively documents the strict border controls, increased costs of crossing borders as well as migrants' reliance on smuggling networks and their experiences of violence along the journey (Collyer 2010; HRW 2014; MSF 2013). Reports and research have revealed that beating, robbery and rape by smugglers and bandits start before migrants arrive in Morocco (Cherti and Grant 2013; HRW 2014). Most migrants entering from the land borders arrive physically and psychologically exhausted after long journeys that may take anywhere from months to several years depending on one's resources (interview with MSF and Terre des Hommes, Rabat, April 2012). The use of coercion in the form of push-backs, removal to the border, and physical abuse define migrants' experiences of the post-entry period and work as mechanisms to push migrants away from the EU borders into urban areas of Morocco. In the Moroccan context, these strict control practices are not limited to the areas bordering the EU. Migrants' experiences of deportability, in terms of their removal to non-EU frontal zones, are not only seen as a possibility, but are a part of their daily reality. Deportability defines the experiences of those in rural areas who are waiting for opportunities to cross the border as well as those who are semi-settled in urban areas. Until the September 2013 reform initiative, commonly reported aspects of migration controls in Morocco included difficulties with mobility after entering Morocco, deportation practices between the EU and Algerian borders and police raids in urban settings (GADEM et al. 2013; HRW 2014). These practices reveal the coercion inherent in what is called 'external dimensions of EU migration policies' and show that the borders of Fortress Europe start long before migrants reach the actual EU borders. Migrants' experiences of deportability at different stages of their journey in Morocco, such as illegal entry, the post-entry journey near the EU border or entry, and settlement into the urban centres constitute major exclusionary mechanisms that make migrants' incorporation into the society increasingly challenging.

Deportability at the borderlands

Despite a relatively liberal visa regime that allows passport holders from several African countries, including Algeria, Congo-Brazzaville, Guinea, Ivory Coast, Libya, Mali, Niger, Senegal, and Tunisia, to enter Morocco legally with a renewable stamp, a significant number of migrants with no passports, who had to flee dire economic and political conditions in their countries of origin, enter Morocco through human smuggling at the Algerian-Moroccan border, officially closed since 1994 because of the conflict over Western Sahara. Oujda, the city situated at the Algerian border, is the main entry point, especially for those who enter without passports. By contrast, migrants with valid documents use the Southern Morocco-Mauritanian border (GADEM et al. 2014: 8-9). From Oujda, those with resources (i.e. money and connections) immediately look for ways to leave for Europe. Others look for opportunities to move to urban centres such as Rabat, Casablanca, and Tangier,[1] where, according to previous research, they are stranded for around two to three years to collect the money needed to move forward to Europe (AMERM 2008). Migrants typically hide in the forests on the outskirts of Tangier and Nador, living in ad hoc camps, while they attempt to cross European borders without documents. There is evidence that controls along the EU border are stricter and more violent in comparison to those at the eastern and southern borders of the country (Migreurope 2006: 11; MFS 2013). Humanitarian agencies identified the rural areas around the city of Nador as the most difficult areas to operate.

As discussed in Chapter 2, Morocco's non-EU borders are not equally equipped with security measures, thus they are more permeable. It is relatively easier and less costly to cross the border between Algeria and Morocco, although it can only be done with the help of smuggling networks. Edith, a 52-year-old woman from the Democratic Republic of Congo (DRC) explained that it is not the crossing of the border into Morocco that is challenging, but the subsequent phase: 'At the borders, they know that we are poor. You pay, but not so much – 50, 100 or 200. This is already too much [for us].' The permeability of the non-EU borders has given rise to different forms of exclusion in the post-entry phase of migration, especially for those entering without documents through the Algerian border. Once in Oujda, it is difficult to exit the city, either to go to big cities such as Rabat or

---

1   Reportedly, Tangier was deserted in 2005 after Ceuta events. However, in the last couple of years, urban migrants have started to settle there again.

Casablanca or to go to the north, near Tangier or Nador, to try to cross into Europe. Most migrants and NGOs operating in the field have underscored that, especially before 2014, migrants had limited mobility after arriving in Oujda. Authorities closely control the city centre and surroundings. Unlike Rabat or Casablanca, it is almost impossible for a foreigner without legal documents to rent a house and/or work in the informal market in Oudja. The police closely monitor the informal settlements in Oujda's forest, near the border, and around the university, and there have been arrests and raids that have destroyed informal camps in the rural areas, pushing migrants back to the Algerian border. 'The police intervene at 4 am in the morning and set fire to the plastic tents,' remarked an NGO worker operating in Oujda, confirming that removal from the forest has become a regular practice since 2006 (author interview, Oujda, September 2012).

In addition to the coercive practices and removals to the border, the major reason migrants are stranded in the forest in Oujda is that foreigners without legal papers are not allowed to leave the city of Oujda by regular train or bus. Note that the Oujda train station was the only place in Morocco where my passport was checked before buying a train ticket to Rabat. As migrants cross the border illegally, they lack the necessary papers. Therefore, they are also denied access to travel to other parts of the country. This situation of de facto denied entry leaves individual migrants stranded in Oujda and renders them more dependent on smuggling networks not only to reach European borders, but also to reach bigger cities such as Casablanca or Rabat.

> Moroccan authorities have established a system of blockage to prevent exit from Oujda, by all means of transport. For instance, they have established police controls in the station. They ask for papers when they see a black person. The same is true for bus stations, for big taxis ranks. That means they have put in place a system of blockage for migrants entering and exiting Oujda (author interview with an NGO, Oujda, September 2012).

The Morocco-Algerian border near Oujda is also the exit and re-entry area for migrants apprehended by the police, either near the border or in urban neighbourhoods, where they are pushed to the Algerian border and re-enter Morocco. Removing these migrants, who are apprehended in irregular situations, to the Algerian border creates a cycle of immobility. Every time migrants are caught without documents, they are deported to the Algerian border near Oujda and walk back to the informal settlements around the city, where they are blocked again. An NGO worker based in

Tangier explained the process of removal to the border and the re-entry as follows:

> In Oujda, they spend 2 to 3 days in the police station. [After removal to the border] they have to walk around 80km to arrive in the city. After Oujda, you need to find a connection to buy a ticket for the bus. If an African student buys the ticket, migrants can escape control. There is no major control after the bus leaves (author interview, Tangier, April 2012).

To overcome this blockage and de facto refusal of entry, most migrants pay to acquire forged papers after entry, based on the knowledge that an identity may protect them from deportation and enable their access to other cities. These papers can help to buy transportation tickets with a fake or borrowed student identity or with a forged asylum application. The cost of fake papers and the journey to big cities varies. Naima, from Central African Republic, needed fake papers to move to the border after entering Morocco:

> As there are controls, you need to have papers. There are people doing fake identities to allow you to get out. These people will also buy you tickets for the bus, train, etc. Like this, they put us in a train and we came here. There are always people you pay, they give you papers and fake identities. It depends on the individual, some people pay 500 Moroccan Dirham (MAD), others 1000 MAD.[2]

The interesting point here is that papers are not only essential for crossing borders, as is widely studied in the literature on human smuggling, but are critical for one's movement within the country after crossing the border without documents. There are several implications of these practices of denial of entry in terms of the production of migrant illegality, migrant incorporation, and access to rights. Because of this system of blockage, migrants are immobilized and illegalized upon their entry into Morocco. Their right to enter and stay within a safe territory, as asylum seekers or as persons who cannot be deported because of their need for protection, as stated in the law 02-03, is denied. As UNHCR does not have an office in Oujda, the access to asylum after entering the territory is not possible. Potential asylum seekers are expected to reach the UNHCR office in Rabat. As emphasized by the informant from the Moroccan Organization for Human Rights (OMDH): 'In Morocco, there are many refugees who are not

2   1 Euro was around 10 MAD at the time.

recognized because they were not able to come to Rabat and apply for asylum' (author interview, Rabat, April 2012). As the implementing partner of UNHCR, OMDH occasionally accompanies migrants willing to apply for asylum from Oujda to Rabat, but this only applies to exceptional cases. In this sense, migrant illegality at the border is reinforced through the denial of access to asylum. Indeed, most asylum seekers face the risk of deportation before they even become an applicant. Fake identities may protect migrants from deportation, but simultaneously increase their dependence on criminal networks. From a legal perspective, by forging papers, irregular migrants, including the potential refugees among them, become foreigners engaged in criminal activities.

As was apparent in the narrative of André, an asylum seeker whose story is briefly presented at the beginning of the book, the strict border controls and coercive practices make entry to Europe riskier and costlier, and push migrants intending to cross to Europe out of border areas and into urban neighbourhoods. Having experienced the hardship of life in the forest area, several migrants interviewed moved to urban centres such as Rabat, Casablanca, and Tangier, where they looked for opportunities to collect money and ways to make progress. In other words, strict controls, harsh living conditions, and removal practices along the EU border create a situation where even the most determined migrants may change their minds or at least settle in urban areas until they find the next opportunity to reach EU borders. Naima, a 29-year-old woman from the Central African Republic, left her husband and two children many years ago and has been travelling alone. She arrived in Oujda after a long journey, passing through Cameroon, Nigeria, Niger, and Algeria with the intention to cross to Spain. She was advised to go to Rabat and apply for asylum after an unwanted pregnancy:

> Upon my arrival, I left for the forest to attempt the journey. We were settled in the forest. After, we attempted, attempted, we were drowned in the water from a small dinghy.. We were stopped. We were sent back. You sleep in the camps. Men go to search for water, the food... there were other women and men. We were in groups. Men and women were sleeping in different areas. Some people were going to the city to search for food. It was a long walk, sometimes in the dark. Sometimes you find tomatoes, not in good condition. Then, we go to "attack". We call this attack. How many people? It depends on the zodiac, if it is small, 15 people.'

When I met Naima, she was expecting a baby as a result of an unwanted pregnancy. 'I was in Nador.' She explained, 'We tried, it did not work. Then,

I was raped. There was pregnancy. There was a brother there, with his wife. Together, we came here.'

This practice of pushing migrants from the EU border to cities reveals that migrants whose primary motivation is to move to Europe spend enough time in Morocco to become de facto members of society and, at times, political actors claiming recognition. Most migrants, seen as being in transit by policymakers and practitioners, are semi-settled in urban contexts along with other migrants in irregular situations who have never attempted to cross the border. Indeed, most association and community leaders have experienced the practices and living conditions along the border. André's story illustrates how migrants' experiences of exclusion at the EU border may translate into political activism in the post-entry period.

Deportability in urban life

Given the hardship at border areas, most migrants decide or are forced to move to urban settings. However, moving to big cities only partially provides protection from deportation practices. Raids by the police in urban neighbourhoods have been part of the daily experience that further marginalizes migrants in their social and economic life, revealing the thin line between deportability as a possibility and deportation as a reality. As discussed in Chapter 2, the production of migrant illegality renders migrants an irregular legal status, as subjects deportable by the state. It has been emphasized that it is the possibility of deportation, rather than its actual realization that makes migrants docile subjects and exploitable workers (see Calavita 2005; Peutz and De Genova 2010: 14; Garcés-Mascareñas 2012). Conversely, in the Moroccan case, deportation has been practiced, until very recently, at the heart of the national territory. Such practices have made deportability a part of the daily experience of illegality, rather than a mere possibility.

It has not been possible to collect data on the frequency of police raids in urban settings, but the practices of removal to the border have changed over time. Informants from civil society organizations drew attention to the unpredictability of the timing and frequency of raids, but also to changes and improvements over the years. Consequently, migrants' experiences of deportability have been subject to change and differ from one group to another. One common point is that the situation is not as bad, at least in urban settings, as it was in 2005 and 2006. Moussa, a migrant from Guinea who has been settled in Morocco since 2002, after trying to cross for several months when he first arrived, explains the changing conditions of deportability over the years: 'Before we could not go out. They [migrants] were hiding in the forest, in [safe] houses. There were a lot of raids. Great change, it is for the better [...] With police it has changed, it is totally better. You see Africans working in construction with Moroccans.' While deportation continues to be a reality, one particular way deportation practices have changed concerns the treatment of groups such as women, minors, and asylum seekers, who are protected by the law. An NGO worker based in Rabat observed:

> I think there are always deportations. It does not change [...]. The deportation of pregnant women has decreased, especially in Rabat and in Casa. In Oujda or Nador, it might happen if you are arrested. In Rabat, Casa, women with babies are not stopped. For men, it is possible, always there are deportations. Before, they were arresting pregnant women. It is even against the law 02/03. (Author interview, Rabat, April 2012)

Not only in law, but also in practice, some groups are defined as less illegal, more legitimate, and hence less prone to deportation than others. Due to the widespread belief that the police would not touch women, being pregnant or travelling with children have become ways for young women to avoid the danger of deportation. Thus, like genuine or forged papers, pregnancies and small babies may serve the function of countering the danger of deportation (Kastner 2010: 22). There is also a widespread belief that babies enable easier access to legal status once the migrants cross into Spain. This is why they are commonly called 'visa babies' or 'protection babies'. One should not forget that pregnancy may be an unintended result of consensual sexual relations or sexual relations during the journey (Kastner 2010). On the other hand, once the project to cross to Europe fails, pregnant women or single women with small children, albeit free from the daily experience of deportability, constitute the most vulnerable group in terms of their participation in economic life, as discussed in the next sections. Naima, for instance, was not sent to Oujda after she was apprehended at the border because she was pregnant; instead, she was sent to Rabat. When I met her in May 2014, Naima was eight months pregnant, unemployed, and hopeless about the future.

Among several English- or French-speaking communities, Senegalese, who can enter the country with a valid passport, are known to be less subject to deportation. More generally, migrants who have a passport with a valid entry but have overstayed their visa are less prone to deportation than those without a passport. Among others, Jules, from the DRC, drew attention to changing practices of deportation. He noted that, previously, everyone was deported but that 'since approximately 2009, if you have a passport, even if it is expired, they will let you go.' Hence, the possession of papers, even if they are not fully in line with immigration laws, provides a degree of protection from deportation. Overstayers in the urban setting are seen as less problematic, as they are considered economic migrants, in Morocco as well as elsewhere. The possession of certain papers protects migrants from deportation, especially those from countries that can travel to Morocco without a visa. Meanwhile, migrants with legal entry are aware of deportation practices, and they are cautious in their relations with the police. Oumar, a 22-year-old Guinean man, came to Morocco by airplane to pursue a career as a football player. Oumar himself was not interested in clandestine migration, but had witnessed 'brothers' being taken to Oujda. Although Oumar had overstayed his three-month visa, stamped on his passport upon entry, he did not feel subject to deportation:

> There are raids. They send you to the border. When you take a room, they will take you out, call the police. I have seen it myself [...] To Spain, clandestine, no! If I were not a football player [...]. My objective is to play football, I cannot become a star [and be] clandestine. [...] When I am in a club, the club will ask for a residence permit for me. Even if your stamp is finished, the police will leave you alone because you have come legally. I was never stopped and asked for papers. I have never spoken to a policeman either.

Despite the diversity in migrant experiences and perceptions of deportability, practices of removal to the border are the most heavily criticized aspect of immigration controls in Morocco. Most NGOs have called for proper implementation of the national law, with respect to international conventions signed by the Moroccan state. Violations of national and international laws by security forces triggered widespread critiques by international and Moroccan civil society and migrants' associations (see e.g. AMDH Oujda 2012; AMDH 2012; MSF 2013). As explained in more detail in Section 3.4, such violations have also provided grounds for migrant mobilization.

After the King's speech

These critiques and recommendations led to a paradigmatic change in Moroccan immigration policies, initiated by King Mohammed VI, as explained in Chapter 3. Removal from urban areas to the border stopped in the aftermath of the royal discourse presented by the King in September 2013. Regarding the continuation of the removals from the EU borders to Oujda, there have been demands by NGOs to stop deportations (Chaudier, 2013). The response of the Moroccan state was not to stop removals completely. NGO representatives and officials confirmed that there were no more removals to the Algerian border, but displacement to Rabat instead. Oujda was reported to be calm as of May 2014 (author interview with IOM, Rabat, March 2014). An official from the Ministry in Charge of Moroccans Abroad and Migration Affairs, charged with immigration issues as of October 2013, responded to the question that I hesitantly asked on deportation practices at the border by first saying 'no more taboos' and confirmed the displacements: 'We have been first to say that there is violation. People are being taken [from the border] to Rabat for integration. It is symbolic.'

How can this new practice of displacement to the cities be interpreted in terms of migrant illegality produced within an international context? As discussed in Chapter 2, there has been a rupture in the Moroccan immigration policy framework. Meanwhile, the border securitization efforts by both Morocco and Spain reveal a continuity in the way EU borders are protected. Morocco remains 'the gendarmerie of the EU', and migrants and smuggling networks continue to alter their tactics of entry. As explained in the previous chapter, since 2013, 'attacks' by migrants have become much more organized, in the sense that migrants now gather in considerable numbers and organize a joint attempt at entry. André, like other migrant activists, has been following very closely what is happening at the EU borders in terms of casualties and success stories:

> Attacks started in 2013. Every year, things change in the forest. As Europeans reflect on raising the barriers, we sub-Saharans also reflect on the tactics on how to get to Europe. If, for example, we are 800, we attack the barrier, 150-200 can enter. Even if the others cannot enter, it is the price to pay.

This situation reveals that the new Moroccan policy for regularization has not changed human insecurities and illegalization stemming from the EU border policies. In terms of migrant incorporation, the practice of pushing migrants towards cities shows that the introduction of a new policy approach resulted in migrants being more welcome to remain within the country, as long as they stay away from the EU borders.

Given the difficulty of crossing into Europe and the conditions of life near border areas, most migrants intending to pass into the EU reach big cities such as Rabat when their project to cross is jeopardized. In the urban setting, so-called transit migrants mingle with other migrant groups, including migrants with legal status, asylum seekers, recognized refugees, overstayers, and undocumented migrants with no intention to cross. In this sense, it is difficult to distinguish migrants in terms of their (alleged) aspirations to go to Europe and their experiences in the urban space. While keeping this observation in mind, in the next section, we shift the focus from state practices that reinforce illegality to the interaction of migrant illegality with existing economic and social conditions.

3.2   Illegality in (semi-)settlement

Settling into violent neighbourhoods

Mama, a 52-year-old asylum seeker, separated from her husband and, along with her biological brother, Jean-Baptiste, fled the civil unrest after the presidential elections in 2010 in Ivory Coast. After staying in refugee camps in Ghana and Togo, they chose to continue their journey to Morocco. Crossing via the southern border with Mauritania, they arrived in Rabat by train.

> After the first night in a hotel in the city centre, the reception man told us to go to [the name of a poor neighbourhood of Rabat] to meet other Ivoirians. We took a white taxi, paid 15 MAD.[3] There were a lot of black people in the neighbourhood. The first black person we talked to knew a girl from Ivory Coast. We were looking for a place to stay. She said she had a cousin, she lived with her boyfriend and they have a big room. As they work during the day, the room is available. We stayed there one month. Then, we went to Caritas [a charity organization of the Catholic Church]. They helped us to find a house. Caritas gives money with the condition that we find a house ourselves, first. So, we looked for a house. We looked from day to night. [...] Then, by coincidence, we met a Senegalese man, a man that I had made acquaintance with in Togo, in the refugee camp. This is how we have found the current house.

Arriving in urban settings, most migrants rely on more experienced migrants to get housing. Edith, a woman in her 50s, from DRC, came to Rabat alone after passing through Oujda. She was later joined by her 'sister' Maria and her five children, who she knew from the DRC. She had left her country

---

3   A white taxi, also called a *grand taxi* ('big taxi'), is commonly used as public transportation. Similar to public buses, they have both a fixed itinerary and prices.

due to economic hardship and because of the conflict that was taking place. She came to Rabat after years of travelling in African countries. She was happy to finally be in a safe country: 'Here, we suffer but there is security, this is what is important in life.' Edith admits that only in Morocco she has felt any solidarity among Africans:

> – We are Africans, I am not racist but it is true. When we arrive in a place, we look for black people, 'excuse me I have just arrived I do not have a place to stay' then they let you in. Even me, when I arrived I was accommodated. It is for couple of months until you find something.
> – Is it with Congolese or even other nationalities?
> – Congolese but also other nationalities. In Africa no, but as we are here, if you are black it does not matter, Ivorian, Congolese, they might help you. It is for a couple of months then you organize yourself and you look for your family. For example, how I left the country, there was a woman who gave me her number. I asked around until I find her and she gave me a place to stay.

In the absence of access to the formal right to stay, most migrants arriving in urban areas rent a house, or rather a room in an apartment, without a contract and in *quartiers populaires*, poorer neighbourhoods of big cities. Finding accommodation without legal papers is possible as long as migrants are ready to pay the price. It is common practice among landlords to ask migrants to pay higher prices than locals. In other words, they are integrated into the housing market by paying a higher price for their integration, as suggested by Cvajner and Sciortino (2010). But, often, there is also an additional price to pay for this informal integration, in terms of violence and opportunistic forms of abuse. In my interviews and informal conversations with newly arrived migrants, they mentioned the *prix de l'integration* ('integration price') that must be paid for settling.

The housing available to irregular migrants is usually in poorer areas where neighbourhood violence is widespread. There, migrants have become targets of aggression and petty crime. 'Even in Rabat, there are neighbourhoods we do not go in the dark. [She cites names of several neighbourhoods]. You cannot walk in the street. If you do, Moroccans will assault you, hurt you, and even kill you if you do not have change. People are stabbed,' says Amadou, a 26-year-old man from Senegal. Similarly, Sunny from Nigeria shows the knife scar he has on his arm: 'Big knife. He did not ask anything. He had a problem. It is because I am black. If you go to 'the office' [he refers to a meeting place for Igbo men], many people have injuries like that.' A lack of papers, forced settlement in poor neighbourhoods, and the lack of protection are interlinked in migrants' experiences of illegality. Migrants in irregular situations do not have access to proper housing because of their lack of papers and lack of financial means. In other words, they are only admitted into poor neighbourhoods with high crime rates. Note that African students with legal papers also live in the same neighbourhoods known to be dangerous because these are the only areas where they can afford a house. African migrants, regardless of their legal status, are more subject to this kind of violence because of their colour.[4] 'These are young Moroccans, 18-25 years old. When they smoke weed and they see you in a corner, they say "mobile phone and money", take out the knife. It is like this,' explains André. Those without legal status face further exclusion, as they also suffer from lack of access to legal protection and services. Because of fear of deportation, as explained in the previous section, most migrants who are subject to aggression are reluctant to go to the police. Some are even reluctant to go to hospitals, knowing that they may not be admitted or will have to pay high fees. Maya, a young activist from Guinea, explains that the neighbourhood violence and the lack of protection she has experienced first-hand led her to join associations:

> There are things happening, it makes me cry. This is why I do not go out that often, and when I do, I go back before 8 pm. I am scared of walking on the road, I meet them [young Moroccans in the neighbourhood] by the road, they do everything and they are not scared of their parents. They do bad things. Do you understand? Somebody was attacked, almost killed, he was robbed. When I heard this, I was disillusioned. He went to hospital but he was not admitted, not touched because he did not have papers. When I learnt about this, I was very very..., I think it is beyond limits. What if he had died that day, because he does not have papers.

It is common for migrants in Morocco and other contexts who lack legal status and financial and cultural capital to live in disadvantaged neighbourhoods and be subject to the clandestine activities and violence that characterize these areas. However, as implied here and further explained in the discussion of mobilization, what is interesting in the Moroccan case is that neighbourhood violence has been one of the exclusionary mechanisms

---

4   The media frequently reports on violent clashes against migrants, see for instance, *Un Séné-galais tué à Tanger après des heurts entre migrants et Marocains* ('A Senegalese killed in Tangier after clashes between migrants and Moroccans'). *telquel.ma*, 01.09.2014. Retrieved 29.03.2015 from http://telquel.ma/2014/09/01/senegalais-tue-tanger-apres-heurts-migrants-marocains_1414696

motivating sub-Saharan migrants to join together under associations. Regarding their association based in one of the most violent neighbourhoods of Rabat, André articulates: 'We mobilize at the moment. In Takadoum[5] you cannot stay calm. You need to be a lion to live there.' Street as well as police violence has been an important catalyst for mobilization. However, it is also a factor impeding migrants' presence in the public sphere. For many, going to community meetings is impossible because of widespread violence. Mama explained to me that she could not attend meetings of the Ivorian association in Takadoum, although she wanted to, because the meetings were late in the evening, and the neighbourhood was dangerous at night. Despite the high rates migrant pay for a place to stay and widespread neighbourhood violence, several informants stressed that the real challenge for migrant incorporation into society is finding a job.

'The problem is work'

Moussa (56, from Guinea), arrived Morocco in 2002, after losing his business and getting 'fooled by his commerce partners.' He travelled with his passport to Morocco and kept looking for ways to cross to Europe, clandestinely. 'Before, it was easier to get into Melilla and into Ceuta. I tried to cross the barriers several times. 4-5 times, many more times. I spent two years in the forest. There are intermediaries. They make money for helping you to pass. [...] We used to leave our passports in the hotel in Tangier,' he says, with the idea of keeping his passport in a secure place in case he does not successfully cross. After several attempts, Moussa came to Rabat where he found daily jobs through his Guinean connections and met his future Moroccan wife. Settled in Morocco for nearly ten years, Moussa has been actively volunteering in a sub-Saharan migrants' association since 2010. At the time of the interview, he had a pending application for Moroccan citizenship. Despite his legal status, Moussa thinks that economic exclusion is the most challenging aspect of life in Morocco. 'When you come, you stay with your friends. Brothers help you until you stand on your feet. The accommodation is not the problem, the problem is work.'

---

5   I have kept the name of the neighbourhood Takadoum, as it is widely referred to in national and international news as an unsafe neighbourhood inhabited by migrants from sub-Saharan countries. See for instance, African Migrants in Morocco Tell of Abuse. *New York Times*, 28.11.2012. Retrieved 29.03.2015 from http://www.nytimes.com/2012/11/29/world/middleeast/african-migrants-in-morocco-tell-of-abuse.html?pagewanted=all&_r=0

A clear relationship has been built, in the literature, between migrant deportability as 'bare life' and illegal migrants supplying cheap labour to the economy (Peutz and De Genova 2010: 14). In this section, the discussion specifically focuses on how migrant illegality does not necessarily translate into economic incorporation into the informal labour market in the Moroccan context, in particular looking at the context of Rabat where most migrants interviewed are based. The structure of the economy and of the labour market only enables marginal participation by migrants. The lack of labour market opportunities has been the major source of frustration referred to in migrants' experiences of incorporation. 'There is no work for us in Morocco' is a common expression of this frustration. As explored in Chapter 2, the production of migrant illegality in Morocco is linked to external pressure applied by the EU to stop irregular border crossings, rather than to Morocco becoming an attractive destination for migrants from the wider region who are seeking employment opportunities. Consequently, the marginalization in the labour market is an indirect result of the international context producing migrant illegality.

Yet, it would be unfair to conclude that the economic incorporation of migrants is characterized by total exclusion. The labour market in Morocco, and more specifically in the context of Rabat, provides certain opportunities that enable migrants to survive. However, the difficulty of finding a regular job persists. Earlier research has revealed that most men work in the construction sector and, to a lesser extent, in restaurants, and sometimes they trade in petty commodities (Pickerill 2011; AMERM 2008). Employment opportunities for women are even more limited. The widespread informal employment sector in Morocco increases vulnerabilities, and migrants always face the risk of being underpaid or not being paid at all (Alioua 2008). Migrant economic incorporation is characterized by employment in certain niches of the economy as well as very marginal economic activities such as begging in the street and sex work.

Niches in the labour market, such as domestic work and call centres, provide opportunities for regular employment to irregular migrants who fit the profiles required by the employers. Gaining access to legal status through work is possible. However, as the procedure is costly and bureaucratic, the majority of migrants work without the necessary documents, either because they find it unnecessarily costly, or their employers are reluctant to provide them. Middle-class Moroccan families employ migrant women as live-in domestics. Senegalese and Filipina women are known to provide domestic work for upper-middle-class families and expats.

As a result of a 1965 convention between Senegal and Morocco, which grants citizens of both countries free circulation and access to their labour markets (DEMIG database 2014), Senegalese do not need a visa to enter Morocco, and, moreover, can get renewable residence permits and a residence permits with the purpose of work when they display a valid work contract. A Senegalese domestic worker, Elou, explains that she was afraid of being deported, and she secured a work permit for herself, even though her employer was not willing to do the paperwork for her. In this case, she made a fake contract in return for money: 'You do as if you work for somebody else, it cost me 2000 MAD. I did it as a precaution, so I can go to the police if something happens to me. I will not renew my card [residence permit with the purpose of work], there is no problem concerning mobility.' Having worked in different countries in the Middle East and Southeast Asia as a domestic worker, Amy from the Philippines thinks that regulations in Morocco for domestic workers are quite flexible:

> This place is not that that strict, and they require having residence. You have your passports, and it is ok with them. But when you have to go back to the Philippines you have to go to police station and ask for the clearance, and after that, you can leave this place. [...] You can always come back, this country is open. It is not like other countries, where you cannot come back if you stay illegal.

Angela, another Filipina domestic worker, could not renew her residence permit after running away from her first employers, where she was sexually abused. After changing employers a couple of times in Casablanca, Rabat, and Tangier, she started working for a 'consulate person' from an African country. 'The employers did not want to do the paperwork because they do not want to be seen as employing illegal migrants.'

Call centres are known as a reliable income source, especially for students from African countries. Working part time or full time is a possible income-generating opportunity, particularly for migrants with advanced language skills. However, informal employment persists here too. Yassine, a Senegalese female university student from Dakar, whom I met whilst she was braiding hair in the 'souk' in Casablanca, had come to spend the summer in Morocco and look for employment. Yassine had a bad experience in the call centre while she was doing an internship. The three-month stamp in her visa had expired, and she was not offered a job by the call centre where she had been an intern for a month. Similarly, Maya from Guinea, whose sister is employed in a relatively well-known call centre, has been disillusioned by her experience in call centres and is no longer interested in finding a job in one:

My sister had Moroccan friends who were in call centres. She found the job thanks to them. I myself did internships, two times in Agdal [a residential, chic neighbourhood in the centre of Rabat]. I stopped. I do not have the will to work there [...] Call centres who are known give you contracts. Those who are not known, small ones, do not give contracts, and they employ you if you are ok. Others even if you are ok, they leave you without a contract, in most cases, they thank you very much.

These examples reveal how illegality is produced in the labour market even for those with legal papers and skills. Rather than giving a contract and doing the paperwork, call centres tend to employ young people with or without a valid status as interns. In this sense, the informal character of the labour market serves as a mechanism for reproducing migrant illegality, even for those who are in more privileged situations in terms of the possession of papers and skills. In fact, most migrants with legal entry indicated that they can only legalize their status by enrolling in private schools, due to the difficulty of getting access to legal papers through work. Patrique from Cameroon has become discouraged by his endless efforts to get a residence permit for the purpose of work and complained about the practices of ANAPEC, the Moroccan National Recruitment and Employment Agency.

We need to know how to submit the file. You register with a private institution. With this [...] you can submit your file to the Ministry dealing with residence permits. Or you make a contract. To have a residence permit through work is almost impossible. I have already tried to apply. I sent my file to ANAPEC for a work visa. ANAPEC procedure is very complicated. Once they pass your file to the Ministry of Employment, it is easier to get your permit. It is ANAPEC that is complicated. I have been waiting for one and half years. I am discouraged; I do not want it anymore.

Maya (from Guinea) underscores that her primary motivation to enrol in a private school is to legalize her status:

– To get my residence permit, I want to enrol in an information technology school. I will go to police with the registration document. I need to legalize my status. The registration is approximately 1000 MAD, then it is 800-900 MAD per month. Depending on the school, you are usually asked to pay for the first two months. Then, you follow courses for the first two months.

– Then, you quit?

– If you want to; if you find it interesting, you may go on. It depends on your means, I would like to continue, but it depends on the situation of my family, do you understand?

Similarly, Moussa's son from his first wife in Guinea joined his second family in Rabat in 2012. Moussa explained that enrolling in private education is not only important for the education of his son, but also for securing his resident's permit: 'He is enrolled in a private school. 700 MAD per month to pay for the school. This is a four-year degree. He will have a residence permit as a student.' Then, he adds with a softer voice, 'his father was clandestine, he will not be the same.'

Migrants without access to regular employment work in daily jobs, for example as construction workers or street pedlars. Jules, a migrant with no documents originally from Congo, does petty jobs for a tailor in his neighbourhood and says that 'this is the only thing I can find.' Street peddling has been common, especially among Senegalese or other migrants with a legal entry, who are allegedly less bothered by the police. André articulates the fragile character of unsteady jobs:

> Most young men work in construction, for 80-100 MAD per day. It is not bad if you can work on a regular basis. It is not that they do not want to work. They wait there until late afternoon. The problem is that there is no work. [...] We kept contact with some bosses we already worked for. They call us when there is work. We cannot do anything outside this.

Street pedlars have become more visible after the reform initiative, especially after raids in urban settings stopped. In a symbolic change, as of May 2014, street pedlars selling electronics, mobile phones, cosmetics, and African accessories are now allowed to have stalls along the walls of the Medina of Rabat on the condition that they do not enter the traditional bazaar, the *souk*. Previously, a few stalls were tolerated, later in the evening, close to the central station, where they were occasionally moved on by municipal police. Paul, a street pedlar originally from Cameroon, explains that he is happy now that he can at least open his stall every day and make some money without the fear of deportation. Indeed, he now prefers to do peddle rather than wait for construction work or volunteer for CSOs without proper salaries. He notes: 'We are not allowed in the *Medina* [the old city centre], maybe in six months' time, it will also be possible.'

Figure 3.2   Street pedlars along the main road, next to the walls of the Medina, Rabat

Source: Photo taken by the author, May 2014

Given the scant possibilities in the labour market, regardless of legal status, most migrants, but especially women with children, are only marginally involved. Many women suffer the stigma of being considered a sex worker, and are frequently approached by Moroccans. At the same time, it is also known that many women are forced into sex work in the absence of other possibilities in the labour market. Edith's sister Maria, a young woman with five children under the age of eight, initially told me that she braids hair, before telling me what she really does to make a living:

> What else you can do? I do this job to buy food, if there is no food, they [her children] start to cry [...]. You sleep with Moroccans, they give you 20 MAD, 50 MAD, you are obliged to take it. What else to do? I do this to earn money because people don't have their hair braided every day. Children cry, they go to school. What shall I do?

A number of migrants and asylum seekers, both male and female, enrol in language courses as well as courses for handicrafts, information technology, and media, offered by associations in collaboration with UNHCR. Some explained that they participate in these courses to spend time together and because they can claim transport costs (around one-two euro per day). As previously mentioned, Mama, an asylum seeker from the Ivory Coast, goes there to forget what she has been through and for the transport expenses she gets from UNHCR, which is her only income besides the money she receives from her relatives in Europe: 'Some courses pay well. At least I learn something. If my brother goes there as well, we could at least pay the rent.'

Mama thinks Morocco is taking better care of these women than other countries she has been to in 'Black Africa'. My general observation is that the day care for the children and babies of the participants provided during the courses offer women a break from their maternal duties. However, many of the participants of such courses complained that it is not really possible to turn the skills they gained during the training into income generating activities in the labour market. But despite their limitation in facilitating migrants' incorporation into the labour market, these courses provide important spaces for political socialization. Migrants participate in associative life where associations provide a public space for them to come together and exchange information, as further explained in Section 3.4.

Begging is another marginal economic activity that is widespread among immigrants in urban areas, especially women with children (AMERM 2008). The fact that the police leave these women with children alone gives them a de facto licence to be on the streets. It is believed that English-speaking migrants, who do not speak French or Arabic, are more likely to beg because of the language barrier that further marginalizes them (Pickerill 2011: 411). Fatima and Sunny's stories are illustrative of the motivations behind begging, given the absence of labour market incorporation.

I met Fatima and her baby Moustapha almost every day during my fieldwork in September 2012. Fatima, from Nigeria, had been begging on one of the main streets of Rabat, leading to the central station. Like other women along the street, she was sitting by the pavement, her phone hidden inside her dress, to protect it from thieves. She would say *merci* when I brought her baby some food or milk, but did not talk much. Fatima appreciated that I spoke to her in English, rather than French or *Darija* (Moroccan Arabic) and let me hold her baby. Fatima thought that there was nothing for her to do in Morocco, and she wanted to save enough money, 1000 euro, to cross to Spain. She would leave her shared room in a poor neighbourhood of Rabat, which cost her 800 MAD (80 euro) a month, and bring her baby to the city centre to beg. She also went to beg near mosques in chic neighbourhoods of Rabat, especially after Friday prayers. She had the baby when she was in Oujda. The father named the baby and also gave her a Muslim name before he left for Libya. Once, I talked to her about an association giving free courses for migrant women and compensating any transportation costs. 'You can leave the baby and have fresh air for a while,' I said. Although she was tempted by the idea and seemed to be considering joining me, she stopped for a second, asked 'how much money would this be?' and gave up on the idea having decided that she could make much more by begging.

Sunny, a 37-year-old migrant from Nigeria had been in Rabat for six months. After spending years in different African countries, he entered Morocco through Oujda with the initial aim of looking for a job. He lives in the basement of a building in a poor neighbourhood of Rabat. While the sanitary conditions within the house were poor, he had a tidy and clean room with a TV and nice clothes. It was extremely difficult for Sunny to find a job in a number of Morocco's cities. 'Work is my problem,' he said. 'Here, I go packing, in the second sector. They give me 55 MAD to do cement. It is not even enough to eat.' He describes begging as his current job:

> – I survive by beg[ging]. Yes, it is true. Sometimes I go with them. Some-times I go to Casa. I go there and stay 4-5 days and I come back; this is how I manage. I ask people to give me money. In Casa, they pay more.
> – How much money do you make a day?
> – Sometimes you are lucky you, make 50 MAD, sometimes a man gives you 100 MAD. If you are lucky, a man sees you and gives you 200. It happened to me this year, during fasting, Ramadan. A woman gave me an envelope; I did not know what is inside. [...]I opened it later and found 100 MAD inside. Some people give me 1 MAD, some brown coins. It depends, some 100 MAD.

Later in the interview, while discussing his experience with the police, Sunny took out a piece of paper that he kept in the pocket of his leather jacket, showing his pending asylum application. While he had been interviewed by UNHCR, he did not seem curious about the outcome, knowing that very few Nigerians are granted refugee status.[6] He uses the asylum paper to avoid deportation: 'I go to Casa. I go with a blanket and spend the night out. If the police stop me, I show them this paper. They gave me a number here. They say, in case the police stops you, you call this number. This is the number.' Given the high rejection rates by UNHCR, an asylum application only offers temporary protection from deportation.

To summarize, despite the availability of a young migrant labour force, Morocco's labour market does not provide many opportunities for

---

6   Out of 215 Nigerian applicants assessed during 2013, only two people had been granted refugee status. (Personal communication with UNHCR Morocco).

migrants with irregular status. Especially when compared to Turkey, as will be discussed in Chapter 4, migrants' lack of access to the labour market, and hence to a regular income, has been the major factor impeding their incorporation into the host society. Even overstayers with passports and legal entry have, at times, found it hard to legalize their status through formal contracts with their employers, although the law does permit this. Irregular migrants have become more dependent on humanitarian aid, and this dependence has become more urgent due to numerous factors, including: securitization along the borders and the pushing of those on their way to Europe into urban centres; the fear of deportation along the border and from urban settlements; neighbourhood violence; marginalization in the labour market; and a lack of access to rights and legal status. These are also factors that push migrants to mobilize among themselves and form communal strategies.

3.3   Access to public healthcare and education

What enables migrant incorporation in the absence of labour market opportunities? What would it take for them to feel they are accepted or even, to a certain extent, integrated? Before moving on to the mechanism of mobilization, this section elaborates on a different aspect of incorporation – bureaucratic incorporation – generally defined as access to fundamental rights and legal status despite restrictive laws (see Marrow 2009; Chauvin and Garcés-Mascareñas 2014). Here, I discuss whether the access to fundamental rights indicates a degree of migrant incorporation despite marginalization in social and economic life. Given the context confronting migrants, both in terms of deportation practices and neighbourhood violence, as discussed in the previous section, access to healthcare becomes a matter of urgency in migrant experiences of illegality.

In the Moroccan case, civil society organizations play a key role in ensuring migrants' access to rights, especially to healthcare and education. This sub-section also indicates the importance of alliances between CSOs and state institutions, as well as between CSOs and migrants' own associations, to ensure migrants' access to fundamental rights. Arguably, in addition to the push factors mentioned above, this close cooperation provided another basis for the mobilization for the rights of irregular migrants in Morocco, as will be explicated in the following section.

Healthcare between formal recognition and bureaucratic incorporation

Regardless of their legal status, all migrants have legal access to free public healthcare, based on a Circular of the Ministry of Health, introduced in 2003. While the main motivation behind this circular has been preventing epidemics and securing public health (MSF 2013: 24), it is still foundational to irregular migrants' access to healthcare. Based on this circular, and the 2011 Law 34-09, which relates to the 'Health System and Offer of Care' and hospitals' internal regulations, the Moroccan legislation recognizes irregular migrants' right to healthcare (MSF 2013: 24; GADEM et al. 2013: 73). However, as documented in several reports, most migrants cannot fully benefit from this right in practice. MSF reports (2010, 2013) have highlighted that migrants in rural settlements refrain from seeking healthcare, as their needs this care often result from coercive border controls. The fear of arrests and deportations discourage them from going to public hospitals (MSF 2013; Moroccan Ministry of Health 2014a). Healthcare access represents a case where even legally recognized rights can only be exercised through the mediation of several stakeholders, such as community leaders or CSOs. The survey conducted by the Ministry of Health confirms that around 42 per cent of migrants surveyed have been subject to violence, and around 10 per cent have been denied access to hospitals (Moroccan Ministry of Health, 2014a). The Moroccan Ministry of Health acknowledges that the access to care is primarily covered through CSOs, informal contracts between CSOs and public health institutions, and through social assistance schemes in certain hospitals (Moroccan Ministry of Health, 2014a). In other words, the legal recognition of migrants' rights to healthcare is only possible through informal incorporation and de facto membership practices.

Although some informants and reports mentioned improvements, very few people interviewed could directly access hospitals. Access to health is managed either through informal community networks or humanitarian organizations. Most migrants interviewed rely on their ethnicity-based fictive kinship networks before seeking institutionalized medical help. They can only access public hospitals through their community contacts and the agency of CSOs. Edith from DRC explains her dependence on 'brothers' and on civil society if she needs to go to the hospital: 'Even here, if you are poor and you get sick you call the chairman, this person takes you to hospital. Women like us, if they do not have the means, they call the chief and the chief calls for help those with means. You give 5 MAD and like this.' André acknowledges the positive change in Rabat's hospitals, which are now more likely to receive patients without asking for passports. Civil society organizations running clinics may provide undocumented migrants with the necessary paperwork that enables them to seek medical care more confidently and also offers migrants certain basic medicines for free.

> Caritas gives you a *carnet* ('paper report') to go to hospital or to a health clinic. You do the consultation and you come back to Caritas. They have a pharmacy. They see if they have the medicine you need. There, they give you the products that do not exceed 100 MAD. These are generics. When you have something more serious, we need to calculate and pay for it.

The healthcare service directly provided by CSOs plays an important role in compensating for the lack of public healthcare, especially in border areas. Given the inadequacy of the Moroccan healthcare system in addressing sexual violence (MSF 2010), the rehabilitation of survivors of sexual violence and women's access to birth clinics have been gender-specific aspects of migrants' access to healthcare (GADEM et al. 2013: 80). A church official who regularly visited hospitals explained that even though African women are admitted into hospitals to give birth, they are usually called *la celibataire* (single) and face discrimination and maltreatment. Therefore, international and Moroccan NGOs collaborate in providing services, but also to create buffer zones that facilitate women's access to sexual health and birth clinics.

While these constitute very important mechanisms that enable irregular migrants' access to public healthcare, they are usually limited in their capacity and may lead to exclusion of certain migrant communities. Angela, a Filipina domestic worker, was pregnant from her Nigerian partner and had stopped working. When she sought help from a charity organization for the birth, Angela was refused, on the grounds that she did not fit the profile. 'This organization is helping people who need help. You are white,' she was told. In fact, she was being rejected due to the 'wrong' colour of her skin, and possibly because she was an economic migrant, an overstayer with a relatively advantaged legal status. Ultimately, through word of mouth, Angela and her husband learned of a female doctor who admits migrants to a public birth clinic. They managed to have the delivery without paying fees. After several attempts, Angela was able to obtain a birth certificate for the baby.

Along with accessing birth clinics, acquiring birth certificates for their new-born babies is another major bureaucratic difficulty confronting women. Birth certificates are crucial for children's juridical existence and their bureaucratic incorporation in later years. A lack of a birth certificate leads to the reproduction of illegality. As discussed in the section on access to education, the birth certificate is a required document for children's enrolment in public schools. Hence, the facilitation of registration of new births is among the recommendations underscored the National Council of Human Rights' (CNDH) report, which has informed Morocco's new im-migration policy (CNDH 2013: 5).

Despite improvements that have increased irregular migrants' access to public hospitals, the medical system in Morocco falls short of meeting the needs of citizens and migrants alike. Migrants' access to healthcare beyond primary consultations remains problematic. On the one hand, the reform initiative in relation to immigration policies envisages a national strategy for improving medical care for irregular migrants who 'should benefit from all the possibilities of medical care in Morocco, with the same entitlement as nationals,' as stated by the Minister of Health in January 2014 (Moroccan Ministry of Health 2014b). On the other hand, the exclusion of foreigners from the new health insurance scheme known as RAMED, introduced in 2012, has raised a number of concerns (MSF 2013; GADEM et al. 2013: 76). The introduction of a centralized electronic system has led to more bureaucratic exclusion, even though the right to medical care is recognized in laws and regulations, as explained above. My follow-up interviews in May 2014 revealed complaints about the new system: 'We are received in the hospitals. It is ok. [Now], [y]ou need a number, also the Moroccans. You cannot receive serious treatment without a number. They changed the system,' explained André. Rosa, a recognized refugee from the DRC, was actually refused treatment in a health clinic because she did not have the necessary documents for electronic registration: 'I understood it was not because I was refugee, there was another Moroccan woman next to me. She did not have the number. She was also refused.' To summarize, irregular migrants' access to healthcare in Morocco is formally recognized. However, migrant illegality has led to different forms of bureaucratic exclusion, rather than bureaucratic incorporation. The situation has been partly ameliorated by the efforts of international and Moroccan civil society, yet it remains to be seen whether the new policy approach will lead to more inclusion with respect to access to healthcare.

Public education: Bureaucratic sabotage and self-exclusion

Along with the international conventions ratified by Morocco, such as the Convention on the Rights of the Child, Article 21 of the Moroccan Constitu-tion acknowledges that the universal right to education is not limited to

Moroccan nationals. Since 2005, the Ministry of the Education has enabled provincial delegations of the Ministry to make decisions concerning the enrolment of children with other nationalities in school.[7] However, in practice, their access to public schools is restricted due to their parents' illegal status, because school registration in Morocco requires a copy of a child's passport or birth certificate. The lack of access to a birth certificate has led to the transmission of illegality from one generation to the next and has deprived children of public education. What is really interesting in the Moroccan case is that despite the exclusionary mechanisms in play due to migration controls, neighbourhood violence, and the situation in the labour market, some children of irregular migrants, albeit modest in numbers, have been able to access free public education as a result of 'bureaucratic sabotage' (Chauvin and Garcés-Mascareñas 2014).

The bureaucratic incorporation of children of migrants of irregular status has been carved out thanks to the growing presence of international and Moroccan civil society networks providing services to irregular migrants. Most CSOs working in the context of Morocco do not distinguish between migrants, asylum seekers, and recognized refugees in the provision of services. In the absence of access to free public education, international and Moroccan humanitarian organizations, at times in cooperation with UNHCR, have provided informal education for children of asylum seekers, refugees, and migrants with no legal status. According to Caritas, a very limited number of children have been accepted into private schools (author interview, Rabat, July 2012). The enrolment of migrant children in public schools is a result of negotiations between UNHCR and the provincial delegation of the Ministry in Rabat. Based on the above-mentioned decision by the Ministry in 2005, the provincial delegation agreed to admit children of refugees and asylum seekers into public schools without the prerequisite birth certificates.

As of the 2009-2010 academic year, UNHCR started providing a list of students to be enrolled in public schools to the provincial delegation in the Rabat-Sale region (see also Qassemy 2014: 13-14). The list is prepared by CSOs that provide informal education to children of recognized refugees, asylum seekers, and irregular migrants. As elsewhere, CSOs in Morocco providing services to UNHCR may, at times, restrict their activities to people under

---

7　La note n°93, du 19 août 2005, portant sur l'inscription au sein des établissements de scolarisation publics des enfants étrangers ('Note no : 93, 19 August 2005, on the enrolment of foreign children in public education institutions') see Qassemy (2014: 46) for the translation of the circular from Arabic to French.

the UNHCR mandate. However, Foundation Orient Occident (FOO), an organization dealing with school enrolment for migrant children, acknowledged that no distinction was made between children of asylum seekers, recognized refugees, and irregular migrants in the preparation and the approval process for school enrolment (author interview, Rabat, September, 2012). In other words, children of migrants without legal status were also included in the list, and UNHCR approved them without distinguishing between people under its mandate and others. Based on the list sent by UNHCR, representatives of the provincial delegations of the Ministry in Rabat wrote to school principals advising them to accept these students without proof of birth certificate (author interview with an official from the provincial delegation of the Ministry of Education, Rabat, July 2012). Accordingly, between 2009 and 2013, 101 migrant children (100 sub-Saharan and one Iraqi) were enrolled in 31 public and five private schools in the Rabat-Salé region (Qassemy 2014:14).

The number of students benefiting from this mechanism of bureaucratic sabotage has remained limited for several reasons. Meanwhile, this rather unofficial practice has provided an opening for more inclusive policies in favour of access to formal schooling for children of migrants in the context of new immigration policies in the post-September 2013 period. A new circular published by the Ministry of Education in October 2013[8] specifically targets access to school for children originating from countries of sub-Saharan Africa and the Sahel Region. Accordingly, identity documents, including birth certificates but now also official documents relating to the status of the parents are required for registration. However, the circular also gives flexibility and discretionary power to regional decision makers by explicating that all equivalent documents showing parents' and children's identity can replace required identity documents. The Ministry also published a note in early 2014, encouraging the integration of children from countries of sub-Saharan Africa and the Sahel Region who are not in formal education, into informal education and into 'second chance' facilities provided by partner associations.[9] Arguably, with the implementation of these two directives, the process would no longer require bureaucratic sabotage with CSOs and UNHCR as intermediaries. Interestingly, both documents explicitly refer to the

---

8   Circular n°13-487, 9 October 2013, concerning the access to education of migrant children from the sub-Saharan and Sahel regions (see Qassemy (2014: 47) for the French version of the circular)

9   Ministerial note n°487-13, 9.10.2013.

constitutional principle of the right to education, to principles of international conventions as well as to the new national immigration policy within the context of greater cooperation and solidarity with people of the sub-Saharan and Sahel regions. In this sense, this transition from bureaucratic incorporation into formal recognition of children of irregular migrants as legitimate members of the society is perceived as part of a wider regional policy.

Despite the availability of bureaucratic incorporation and the recent formal recognition of the universal right to free public education, the access to education for children of irregular migrants is further complicated by a process of self-exclusion. The limitations of bureaucratic incorporation are inextricable from parents' unwillingness to enrol their children in Moroccan schools. Reportedly, several families have disappeared after the enrolment of their children, as they prefer to stay in the camps near the EU borders (author interview, Rabat, September 2012). In other words, children are denied the right to education due not only to exclusionary policies, but also because of their parents' semi-settled situations and their ongoing aspirations to cross into Europe. Women such as Fatima, begging in the streets of Rabat, or Allasane, whom I met with her three children in Tangier while she was looking for a suitable opportunity to cross to Ceuta, are not only economically, socially, and legally excluded, but are less interested in being incorporated into Morocco because of their experiences of exclusion:

> Our situation is far worse than single people. My children do not have birth certificates. How will my children go to school? I want to go to Europe with my kids. There, they can go to school. They are used to French schools. What would they do in Arab schools?

The aspirations to cross to Europe have significantly influenced migrants' reluctance to send their children to school. Rosa, a 42-year-old refugee woman from the DRC, prefers to send her children to private French colleges, rather than public schools, motivated by a belief that French education will help her children after the resettlement she has been waiting almost 20 years for: 'Yes, in a private college, because all schools here are in Arabic. It is always in Arabic, what is she going to do with Arabic. She was so far in Arabic schools and then she said, 'mom, it is not working,' this is why we had to change, so that she could learn some French [...].' As mentioned at the beginning of the chapter, the Moroccan public education curriculum has been one of the most criticized aspects of the Ministerial circular. There is a widespread conviction that the content of public education, which also includes Islamic religious education, is not suitable for Christian children, and those who are not fluent in both French and Arabic face further difficulties (Qassemy 2014: 20; 28). For André, the circular was a failure, because it was initiated without consultation with civil society or migrants' organizations that have a deeper knowledge of the field. Meanwhile, access to public schools constitutes a relevant example of migrant bureaucratic incorporation: the agreement between the Moroccan state and UNHCR for the inclusion of specific groups under the UNCHR mandate, i.e. children of asylum seekers and refugees, was extended to migrants in irregular situations through a series of bureaucratic moves.

The discussion of access to healthcare and public education is also important to show how civil society efforts lead to, albeit de facto, recognition of irregular migrants' rights. Access to healthcare constitutes a case where the formal recognition of a universal right can only be exercised with the mediation of non-state actors. In this sense, the case of health exemplifies bureaucratic exclusion as well as informal forms of inclusion. Minors' access to public education is not only important because it is a fundamental right secured by international conventions, but it also has a symbolic value in terms of migrants and their children being seen as de facto members of the society, despite their illegal status. The access to public education constitutes a case of bureaucratic incorporation, which was subsequently translated into a formal recognition of undocumented migrants' right to free public education. This section has already hinted at growing interconnections and collaboration between migrant communities and civil society enabling migrants' access, albeit marginal, to certain rights. The next section turns our attention to how these interconnections have underpinned communal strategies that migrants embrace in claiming recognition and rights, which are distinctive aspects of the incorporation experience in the Moroccan case.

3.4   Reversing illegality through mobilization

What institutional factors enabled irregular migrants' political incorporation in the Moroccan context? Among the political opportunity structures that are available to migrants, as discussed in Chapter 2, I would suggest that the most important is the simultaneous emergence of Moroccan civil society actors, critical of state policies and practices towards irregular migrants, along with migrant organizations. In the Moroccan context, migrants'

Figure 3.3   **Members of the Democratic Organization of Migrants Workers taking part in a march organized by Moroccan CSOs during pre-COP22 meetings in Tangiers, 24.10.2016**

Photo: ODT

mobilization for rights has become a form of incorporation. In order to explain how the institutionalization of civil society provided opportunity structures for migrants' own mobilization, the following sub-sections discuss the emergence of civil society actors and their main activities. Then, I move to the incorporation of migrants as vocal political actors into these recently emerging institutionalized civil society structures. Institutional and discursive factors have underpinned what I call the 'political incorporation of migrants' in Morocco.

Emergence of civil society networks

As an unintended consequence of the previously discussed events in Ceuta and Melilla, migrants who were forced to move out of rural camps and the city of Tangier in the north have become more visible in big cities. In response to this, there has been a proliferation of international and Moroccan NGOs as advocates of rights and/or providers of humanitarian support to immigrants. NGOs have become important actors, enabling migrants' incorporation and mobilization despite their low capacity and the challenges they face in terms of tense relations with authorities (Collyer et al. 2012: 12). The emergence of civil society working on irregular migration issues has to be contextualized in the wider political and institutional liberalization process, as explained in Chapter 3. Not surprisingly, their visibility in the Moroccan context coincides with the increasing visibility of irregular migration issues. Interestingly, the emergence or expansion of activities by formal international, national, and local CSOs coincided with the emerging politicization of informal sub-Saharan migrants' associations.

Civil society actors working on immigration-related issues in Morocco can be categorized based on their affiliation, activity areas, and relations to state authorities.[10] Concerning affiliation and scope, there are international NGOs such as: MSF (operating in Rabat, Oujda, and, to a lesser extent, in Nador until March 2013); the Catholic Church's charity organization Caritas (with reception centres in Rabat, Casablanca, Tangier); Terre des Hommes; the Protestant International Mutual Aid Committee in Rabat and Oujda; and the French organization CIMADE. Among those older and nationally organized institutions that have developed an interest in immigration as part of their general mandate are: the Moroccan Association for Human Rights (AMDH) and the OMDH; labour unions such as the Democratic Organization of Labor (ODT), so far the only union accepting migrants (with or without legal status) as members since July 2012; organizations operating locally such as the Foundation Orient-Occident (FOO) (in Rabat, with branches in Casablanca and Oujda); ARMID (Association Mediterranean Encounter for Immigration and Development); Pateras de la Vida and CHABAKA in Tangier; ABCDS (Association Beni Znassen Culture Development Solidarité) in Oujda; and AFVIC (Association for Victims of Clandestine Migration and their Families)[11] and GADEM (The Anti-racist Group for the Support and Defence of Foreigners and Migrants) in Rabat. It is also possible to categorize these institutions based on their activities with migrants, asylum seekers, and refugees in Morocco. While humanitarian aid is a priority for some institutions (e.g. Caritas and MSF), others are more preoccupied with legal consultation (e.g. OMDH) and rights advocacy or raising awareness in favour of immigrants (e.g. GADEM, AMDH, and CIMADE). However, in a context where the protection needs of irregular migrants stem from the inadequacy of policies and violent practices (Collyer

---

10   See Table 5, for an overview of NGOs interviewed.
11   No longer active.

2010), most civil society actors approach the situation through a combination of humanitarian aid and advocacy. It is also possible to categorize these civil society actors by their relation to the Moroccan state and various funding bodies (see Jacobs 2012).

Migrants' self-organizations

As discussed in Chapter 1, migrants themselves are incorporated into this emerging civil society network for the rights of irregular migrants, not only as beneficiaries of certain services, but also as active political agents seeking rights. Migrant associations were formed as one of the initial communal strategies for negotiating grievances that stemmed from deportation practices and limited economic opportunities and rights, as explained in the previous sections. The Council of Sub-Saharan Migrants in Morocco (CMSM) and the Collective of Refugees in Morocco were established immediately after the events in Ceuta and Melilla in 2005. The following year, many of the founders of migrant associations collaborated in *Asile Maroc* (Asylum Morocco), jointly organized by the French organization CIMADE and Morocco's AFVIC, together with UNHCR (Alioua 2009). This collaboration was aimed at raising awareness about the issues of asylum and irregular migration. In this sense, interactions between migrant activists and Moroccan or international associations have been strong since the beginning of the mobilization process. As they became more established, increased in number, and collaborated more frequently with Moroccan and transnational civil society actors, sub-Saharan migrant associations increased their visibility and their demands for the fundamental rights of migrants, the regularization of undocumented migrants, and the formal recognition of their associations. The widespread use of the French language among middle-class Moroccans and sub-Saharan migrants, a legacy of French colonialism, facilitated communication between associations and among French-speaking migrant communities, but excluded English-speaking migrant groups. These developments were accompanied by the foundation of several ethnicity-based solidarity associations and African student associations, some of which are recognized by law. Smaller, issue-based migrants' associations such as the Collective of Sub-Saharan Migrants in Morocco (founded in 2010) and ALECMA (Association Lumiere sur l'emigration clandestine au Maghreb) (founded in August 2012) joined later.[12]

---

12   After the regularization campaign, a number of migrant community organizations have found the opportunity to regularize their status.

Denouncing violence against sub-Saharan migrants has been the main motivation for the establishment of solidarity networks, as an ALECMA representative explained:

> There are many sub-Saharans living in Takadoum, it is the hottest neighbourhood in Rabat. [...] this is what motivated us sub-Saharans to come together to create an association, ALECMA. This is to denounce different problems we encounter in the country, then to defend our rights because as migrants, our rights need to be respected, that's it. This is why we regrouped under an association. We started this fight to be recognized [...]. This is related to different acts of aggression. In August 2012, there was a series of attacks. In one week, there were six cases of aggression. This is why we called all sub-Saharans living in Takadoum and we held a massive march. We wanted to be heard. We marched to the police station because a sub-Saharan was seriously injured because of aggression. We marched with the injured to the police station. We passed by consulates, Mali, Ivory Coast, Central Africa. After this march, we had the idea of getting together under an association. (Author interview, Rabat, May 2014)

The use of the word sub-Saharan in the name of associations connotes a common identity and solidarity beyond ethnic, national, and religious differences within the community. In a sense, it is a counter-discursive strategy that opposes the stigmatization of sub-Saharans as 'illegal migrants'. In response to my question about his feelings towards the use of the term sub-Saharan, André replied: 'when I say sub-Saharan, we need to clarify. Moroccans call us Africans. Maybe they are Europeans, I do not know. I am proud of being sub-Saharan [...] I am not bothered about being called black or sub-Saharan.'

Most of the time, migrants refer to their personal background of activism as the main motivation for joining or initiating migrants' organizations. There are others who have gained awareness through a process of political socialization since they arrived in Morocco. An activist from CMSM narrates his personal mobilization story, resulting partly from his activist background and also from his experiences in Morocco. 'I was in an association in my country. After coming here, I went through a training. There was need to s'indigner [revolt]. In 2005, I was in the heart of the events. This gave impetus to my engagement. It was partly what drove my engagement.' Moussa (56, from Guinea) explains his own and other sub-Saharan migrants' involvement with associations as the result of a gradual awareness and reaction to the racist discrimination they face in daily life.

– Since 2005, we created the Council of Migrants. It was the first time there has been an association for the defence of migrants' rights in Morocco. At the beginning I was here but I was not interested. I have been a member for 2-3 years.

– What happened to make you decide to become a member?

– You know, I decided, with brothers, we were organizing things for recently arrived boys. God created the earth for humans to live. Frontiers are not acceptable. The earth does not belong to anyone, it belongs to people. We are fighting for an earth without frontiers. You can live anywhere, regardless of nationality and [there should be] equality for all. The earth is for everyone. This is the aim of our associations.

– Why do you think some migrants are more activist than others?

– It is about communication. We are mobilizing people to rise up for their rights. You should not stay hidden in your house, you need to go out and ask for your rights. You have workers' rights, the right to papers, the right to access to health, the right to liberty. You should not stay in your corner. You need to claim your rights. This is what the Council is trying to do. It is not only for sub-Saharans, it is for all foreigners, Tunisians, Asians. It is true that there is discrimination. The other North Africans [Maghrebins], Asians, Americans, the police does not stop them. There is discrimination.

Although migrants' organizations differ in terms of their internal organization and priorities, they all share a gradual increase in their demands for fundamental rights, for the regularization of migrants, and also the regularization of their associations. It should be noted, only a small minority of immigrants in Morocco are attached to sub-Saharan organizations. Nevertheless, the simultaneous emergence of an international and a Moroccan civil society working on irregular migration has constituted an opportunity for migrants' organizations to set their agenda and raise their claims. Gaining visibility and seeking recognition was possible through collaboration with Moroccan and other international actors in the field. The main axis of collaboration between migrants' organizations and Moroccan and international organizations has been in two areas. These include humanitarian assistance in the field and advocacy activities. Each side admitted its dependency on the other to further its activities and agenda.

The communal strategies of migrants' organizations included direct collaboration with other civil society actors in the field concerning humanitarian aid and legal support, public manifestations and forging formal and informal alliances with Moroccan and transnational actors. It was noted that migrants' organizations are much more efficient in the field, and other associations need them in order to reach the target population. One type of voluntary, and occasionally paid, job that is available to migrants (both irregular and students) is the role of *agent de proximité* (outreach agent) for organizations that conduct research and do humanitarian work in the field. While this may provide an opportunity for the economic incorporation of a small minority of migrants, these relations are not free from tensions. Because the structure of funds that CSOs rely on does not always stretch to remuneration for activists volunteering on the ground, GADEM notes that there are often misunderstandings and frustrations concerning the voluntary work by activists (author interview, Rabat, May 2014). Indeed, several activists have noted a sense of frustration: 'They need to know how to manage people, we are heads of families. Instead of going to Nador, you can do three days of work. You do not do this because you have chosen the road of activism. They exploit us, sub-Saharans. They exploit sub-Saharan activists,' says André.

Despite tensions related to the voluntary services provided by migrants' associations to international and Moroccan CSOs, and the widespread conviction that migrants should be able to speak for themselves, most members of migrants' organizations agree that they rely on Moroccan NGOs to facilitate their public activities. Moroccan NGOs can secure legal permissions for public protests on particular days (Jacobs 2012: 72), such as International Migrants' Day, the anniversary of Ceuta and Melilla events, and International Labour Day. For instance, the Social Forum on Immigration took place in Oujda, in October 2012, with widespread participation from civil society actors from Morocco and from the region. The Collective of sub-Saharan Migrants in Morocco noted that it was AMDH that helped the transfer of migrant participants from Rabat to Oujda, a highly controlled trajectory (author interview, Rabat, September, 2012). As an example of regularization from below (Nyers and Rygiel 2012: 15), Moroccan associations have negotiated with the authorities to ensure the political participation of migrants without legal status. During the preparations of the Social Forum on Immigration, an Oujda-based organization noted that 'the security question was raised in the meetings. We need to ensure the protection of undocumented migrants. We are negotiating with the authorities to receive them here. There is a commission to do this, to facilitate the participation of undocumented people' (author interview, Oujda, September 2012).

The initiative to organize migrant workers under a Moroccan labour union is a concrete example of alliances between migrants' and Moroccan associations for the regularization of migrants. The first step was taken on

Labour Day in 2012, with the announcement of the regularization campaign and a public demonstration. The motto was 'we, also have rights'. The admission of migrant workers into ODT under its new branch, ODT-immigrant workers (ODT-IT), was officially launched with the first congress in July 2012, and hundreds of migrants participated.[13] CMSM played an active role in coordination, together with Marcel Amiyeto,[14] a recognized refugee and the secretary-general of ODT-IT. Amiyeto's narrative on the process of unionization underscores the importance of forging alliances:

> Sub-Saharans are working in the factories, construction, call centres, everywhere, but they are not recognized. When there is an accident, they do not have social security coverage. This is why we intervene to create the union. It took us many negotiations, First all, not all associations agreed with this idea, they did not want the creation of the union. Members of ATMF [Association of Workers from Maghreb in France] encouraged us. We made contacts with ODT and started to reflect on the question and how to make foreign workers members of the union. Before, the internal rules of the union were nor allowing the membership of foreigners.

In this case, migrants' organizations in Morocco forged a transnational alliance with ATMF, a labour union representing migrant workers in France. The interest of the unions in general and the foundation of a migrants' union in particular have been a surprising form of alliance, considering the earlier discussion about how migrant incorporation has largely been defined by economic marginalization. The main contribution of unionization has been the strengthening of alliances for the regularization campaign. CNDH and The Council of the Moroccan Community Living Abroad (CCME) supported ODT-IT to make a formal regularization claim.[15] Both CNDH and CCME are led by Driss El Yazami, a well-known human rights activist in Morocco, appointed by the King as the head of these institutions. These alliances with public figures and key institutions have increased the visibility of irregular

---

13   *L'ODT ouvre ses portes aux travailleurs immigrés* ('ODT opens its doors to migrant workers'), *L'economiste*, 03.07.2012. Retrieved 20.03.2015, from http://www.leconomiste.com/article/896170-l-odt-ouvre-ses-portes-aux-travailleurs-immigr-s

14   I have used his real name, as he is a well-known figure.

15   CCME was established in 2007, and its role was officially recognized in the 2011 Constitution. The Council is constituted of representatives of the Moroccan community abroad, most of whom are appointed by the King himself. See Üstübici, 2015 for an analysis on interaction of emigration and immigration policies in Morocco.

migrants in Morocco. Migrants used their rather limited institutional capacity to reach key institutions, such as CNDH, that are capable of pushing for change in immigration policy. In this sense, even before the publication of their recommendations of the new policy in September 2013, CNDH was a crucial institution for channelling irregular migrants' demands for regularization.

Brothers in arms: What makes alliances possible?

Alliances have been enabled due to the common agenda of protecting and meeting the needs of migrants, along with several additional factors. One major component of the alliances is the common repertoire that Moroccan NGOs and migrants share for legitimizing their alliances and their demands. The relative liberalization of associative life in Morocco, discussed in Chapter 2, has, arguably, allowed Moroccan civil society actors to voice their critique against the state, by relying on discourses of universal human rights, international law, and the rule of law. These principles provided a suitable foundation for raising human rights violations against irregular migrants and asking for regularization. While the security-oriented approach of Law 02/03 is criticized, most NGOs simultaneously urge respecting the protective measures in the law. As an emigration country, Morocco was one of the first nations to sign the 1990 Convention on the Protection of the Rights of All Migrant Workers and Members of Their Families. During the 2000s, the document became a major legal reference for criticizing the treatment of immigrants in Morocco. The responsibilities of Morocco as a signatory of the 1990 UN Convention are continuously underscored in public declarations and meetings. For instance, GADEM (2013) prepared a report on the implementation of the 1990 Convention and implications for the rights of immigrants in Morocco in collaboration with migrant and other Moroccan associations (see GADEM et al. 2013).

Migrant activists' narratives revealed direct references to the democratization process and the ways that migrants situate themselves as progressive actors in this process. One activist from CMSM underscored that Morocco is the first country to have migrants' organizations that do advocacy work, and that this presents an opportunity rather than a threat for the future the country:

> We are doing sit-ins, we are on the TV. They think that we are here to sabotage Morocco, when we criticize state; not really. We are not a threat to Morocco, we are a chance for the country. In terms of associations like

us, in terms of migrants' communities joining together to defend rights, Morocco is the first among all countries in the Maghreb. Honestly, it is a chance for Morocco to respect democratic rights. (Author interview, Rabat, September 2012)

Secondly, these claims are coupled with the observation that the conditions of transit migration have changed. Almost all NGOs and experts that I contacted drew attention to the changing conditions and temporality of transit and underscored that being in transit is no longer a matter of a few weeks as it was ten years ago. An emphasis was placed on how the length of this transit period makes Morocco responsible for the situation of immigrants within its borders, regardless of whether or not they are en route to Europe: 'With AMDH and other associations, we insist on regularization. Some say it is an EU problem, not Moroccan, the EU wants us to regularize people, but they do not want to open their borders. We say it is also a Moroccan problem,' notes ATMF (author interview, Rabat, September 2012). Migrants' associations reject the argument based on the transit status of migrants, stating that the exclusionary policies of the state deny their access to rights and legal status and perpetuate migrants' vulnerabilities in society. This approach also leaves people with no option but to explore the viability of clandestine migration. On the one hand, migrants themselves are explicit about the fact that they are stranded in Morocco because they are unable to reach Europe. On the other hand, they emphasize that migrants are also stuck because policy circles turn a blind eye to their situation and because they are denied the option of staying in Morocco or going back to their countries. As an activist from CMSM put it:

We are undocumented because we are denied documents [...]. You stay for ten years, it is as if you arrived yesterday, you do not take a step. It is heavy for human life, which has a limited time. This is the force that makes people take the sea route. This situation leads people to sad ends, especially if they are running from execution and misery in their own countries. (Author interview, Rabat, September 2012)

Thirdly, alliances are underpinned by common references to sub-Saharans' and Moroccans' shared African identity and shared emigration experience. 'We are all Africans' is a common statement by Moroccan NGOs and activist migrants alike. In public statements, pro-migrants' rights actors display solidarity with 'African brothers', referencing Morocco's African identity.[16] Similarly, King Mohammed VI underscored Morocco's African identity in his royal speech announcing a new approach to immigration policy.[17]

One important repertoire legitimizing civil society interest in immigration issues and the alliances they forged with migrants' associations concerns references to the emigration experience of Morocco,[18] which also reflect the experiences of Moroccan NGOs. AMDH, founded in 1979 after several years working on human rights violations against Moroccans in Europe, reshaped their activities in line with the changing migration scene in the country. They have started to place equal emphasis on human rights violations against immigrants in Morocco: 'We suffered from racism and we are racist against migrants' (author interview, Rabat, September 2012). In this sense, mobilization for the rights of emigrants has influenced attitudes towards immigrants with or without legal status in the country.

References to the emigration of Moroccan families are also used to raise awareness about the vulnerable situation of immigrants in Africa. The statement by the ABDCS reveals how reflecting on the emigration of Moroccans to Europe facilitates communication about the situation of sub-Saharans among Moroccans with families in comparable positions in Europe:

> Sending countries, African countries have a better understanding of the phenomenon of migration. We use this argument when we do awareness-raising in poorer neighbourhoods. When they ask who these people are, what are they doing here. Morocco is not a European country. We say 'think of your brothers in Europe, they are also sans-papier'. There is solidarity despite racist attitudes, we should not forget. (Author interview, Oujda, September 2012)

Taking emigration as a reference point has allowed for the emergence of transnational and at times, in the words of Coutin, for 'unusual alliances' (2011: 302). The use of references to Morocco's emigration history to make

---

16   See for instance, *Non aux violations flagrantes des droits et dignités des frères subsahariens au Maroc*. ('No to brutal violations of the rights and dignities of sub-Saharan brothers in Morocco.'), *atmf.org*, 26.04.2012, Retrieved 29.03.2015, from http://atmf.org/Non-aux-violations-flagrantes-des

17   See for instance, Royal discourse in the occasion of 38th Anniversary of the Green March, 06.11.2013. Retrieved 29.03.2015 from http://www.map.co.ma/fr/discours-messages-sm-le-roi/sm-le-roi-adresse-un-discours-la-nationl%E2%80%99occasion-du-38eme-anniversaire

18   For a detailed analysis of emigration policies and institutions on immigration, see Ustubici, 2015.

a case for migrant rights in Morocco reveals the need to re-think available opportunity structures in transnational terms, as proposed by Però and Solomos (2010: 9). As mentioned, ATMF, the union of workers from Maghreb operating in France, supported the unionization of immigrants in Morocco. For one member, empathy with immigrants in Morocco stems from Moroccans' own experience of their irregular status in Europe: 'It reminds us of our situation in the 1970s. It is natural that we react to this. It is normal. This is why we play the role of advocates' (author interview, Rabat, September 2012). Recently, the CCME, as a quasi-independent state institution, has developed an interest in the situation of irregular migrants. CCME conducted research, in collaboration with the Institute for Public Policy Research in the UK and the Platform for International Cooperation on Undocumented Migrants based in Brussels, on the precarious situation of sub-Saharan irregular migrants in Morocco (see Cherti and Grant 2013). CCME, in collaboration with CNDH, has played a critical role in initiating a new immigration policy in Morocco. Such collaborations and CNDH's ongoing interest in irregular migration in Morocco as a human rights issue, as well as their close relations with migrants' organizations, has increased the visibility of immigrants as political actors. Alongside other geopolitical and foreign policy concerns, migrants' visibility as political actors contesting their illegal status has influenced CNDH's recommendations for a radically new Moroccan immigration policy. Their relative empowerment and increasing concerns over ongoing practices has resulted in an increased momentum in migrant mobilization in the wake of the reform initiative. Moroccan and migrant organizations have continued to organize public protests against rights violations, racist crimes, and attitudes towards migrants.[19] Migrants' associations have found new ways to communicate with the Moroccan government regarding the ongoing regularization campaign, including the regularization of their own informal associations, as narrated at the start of the chapter.

Mobilization for individual mobility

Migrants with no legal status in Morocco have been able to carve out a political space thanks to alliances built with Moroccan and international organizations. The Moroccan case illustrates a rich and underexplored

---

19   See for instance, *Manifestation contre le racisme à Rabat* ('Manifestation against Racism in Rabat'). *yabiladi.com*, 21.09.2014, Retrieved 29.03.2015 from http://www.yabiladi.com/articles/details/29402/manifestation-contre-racisme-rabat

empirical basis for re-investigating the theoretical connection between migrant illegality, incorporation, and mobilization in a context characterized simultaneously by emigration, immigration, and transit. Interestingly, this is also the context in which migrants themselves simultaneously negotiate transit and settlement. In other words, the demands for regularization aimed at the Moroccan government were coupled with more general demands on the freedom of circulation. Predictably, critiques also targeted EU policies restricting the right to asylum and mobility. Many migrant activists were initially motivated by their own experiences of exclusion and their mobilization for rights and legal status in Morocco. Meanwhile, their experiences of mobilization have now been reconciled with their individual projects to go to the EU. In other words, while getting organized, migrants simultaneously continue to explore opportunities to cross to Europe. In this sense, in Morocco, mobilization for the rights of migrants can be a way to acquire social capital that enables migrant activists to travel to Europe legally. Mobilization becomes a means for 'transiting to Europe' but not clandestinely, as the term connotes. For some, transit migration may not have been a motivating factor for coming to Morocco. At this point, the changing meaning of transit needs further analysis from the perspective of migrants. Amadou, a young man, in his mid-20s, from Senegal, for instance, came to Morocco to study, with no interest in migration issues and no prior experience of activism. Living in a poor neighbourhood, also inhabited by migrants with irregular status, and volunteering for some charity organizations, he explains that he has become militant during his stay in Morocco. Amadou later married a French woman that he met through these activist networks. He subsequently moved to France to join his wife and continue his education.

During one of my follow-up visits to Rabat, André told me the story of a community leader I had interviewed but lost contact with: 'He left for France. He left legally. He first went to a forum in Italy. He had the visa. Then, he was invited to France. He left and preferred to stay there. He had finished his post. He left his place to somebody else.' Mentioning another activist I knew, who was invited to a European country and also stayed there, André's tone revealed his appreciation for the success of his peer: 'We are all happy for him, he was a real militant.' Later in our conversation, André said that he was not able to take advantage of similar invitations because he entered Morocco without a passport: 'We always receive invitations, but it is not easy as we do not have passports. We say if they regularize us, we can also make passports and when there are forums around the world. It would allow us to travel in a good way. This is a little bit like that.'

Even those who have not (yet) left Morocco for other destinations reported that they have been empowered by this mobilization process. Maya, from Guinea, a young member of ODT whose studies have been interrupted, but who aspires to resume her studies in Europe, refers to her experience of mobilization for the rights of irregular migrants in Morocco as training. 'Here, it is a form of training. I train myself here, and I see many things happening in different associations. It is knowledge. I tell myself that it is also a school. It is a school.' In other words, for several migrant activists, the mobilization process is a process of incorporation into Moroccan society, but also an opportunity to further their journey as well as a way to give meaning to their semi-settled status In Morocco.

Conclusion: Morocco as a case of political incorporation

The purpose of this chapter has been to reveal the linkages between (i) the production of migrant illegality even before migrants arrive at their alleged final destination; (ii) socio-economic structures that enable and disable migrants' incorporation in new immigration countries; and, (iii) factors that make political activism a viable option for migrants in irregular situations in Morocco. Regarding the mechanisms of control (laws and implementation) and structural factors (the labour market) that produce migrant illegality, Morocco has initially been a case of exclusion at the levels of policy, discourse and practice. Regarding the question of incorporation, Morocco is a crucial case for the study of mobilization for the rights of irregular migrants. The chapter has argued that the interaction between exclusionary practices and other structural, institutional factors has resulted in a particular incorporation style, which I characterize as legal, economic, and social marginalization but political incorporation through mobilization.

The chapter showed the conditions under which the trans-Saharan journey through Morocco to Europe has become a political journey for migrants 'stranded' in Morocco in the post-2005 period. The first three sections focused on migrants' experiences of deportation and labour market participation and their formal and informal access to rights through civil society. The first section confirms the perspectives set out in the literature that suggests that migrant illegality is reproduced through practices of controls and deportation along the borders and in the urban setting, and that the situation reinforces migrants' continuous sense of deportability. This sense of exclusion is exacerbated by marginalization in the labour market and widespread violence in the neighbourhoods where they live, as explained in Section 3.2. Informal solidarity networks comprising migrants and civil society only partially alleviate social exclusion by enabling formal and informal access to fundamental rights and services. The access to healthcare and education are described as two key areas that illustrate how mechanisms of formal and informal bureaucratic inclusion work on the ground. Section 3.3 conceptualized healthcare as an area where the access to a right that is recognized by the state is negotiated through informal practices. It was only after non-state actors applied pressure that public hospitals started to receive immigrants without asking for their legal documents. Access to education is another crucial field where formal access is achieved after a process of informal inclusion and bureaucratic incorporation. Up until the circular on universal access to public education issued by the government in November 2013, the children of irregular migrants could be enrolled in schools via NGOs and UNHCR even in cases where the parents were not asylum seekers or refugees. In contrast with healthcare, the right to education depended on bureaucratic tolerance rather than formal recognition. The Circular on the education of children of migrants formalized the bureaucratic tolerance within a wider context of integration policy.

The fourth section analysed the institutional and discursive contexts that made the political mobilization of irregular migrants a viable option. Migrants with an irregular status in Morocco, animated by their experiences of marginalization, have been able to carve out a political space to claim rights and legal status thanks to alliances with Moroccan and international organizations. The use of a referential framework based on a language of rights, a common African identity, and experiences of emigration reinforced the shared ideational ground of such alliances. Because of this mobilization, migrants with an irregular status gained public visibility and, in turn, were recognized by state authorities.

This aspect of migrants' incorporation as political subjects makes Morocco a crucial case for exploring the link between the literature on migrant illegality and the literature on new social movements, which has paid little attention to new immigration contexts at the periphery of Europe. Political incorporation is also a process through which migrants benefit as individual actors. It is noteworthy that several association leaders found ways to travel to Europe legally through transnational connections built within activist networks.

Despite the limitations on mobilizing for the rights of irregular migrants, the Moroccan case generates interesting empirical and theoretical questions. It remains questionable whether this ongoing political activism in the context of the reform initiative will alleviate migrants' experiences of exclusion and marginalization in social and economic life, the reasons that initially pushed them to mobilize. It is also debatable whether mobilization provides a critical opening for questioning border and membership practices of powerful actors such as the EU or nation-states of the North, or whether it reinforces the international regime of migration controls envisaged by these powerful actors in the first place.

It is worth noting that the mobilization by irregular migrants at the periphery of Europe is more of an exception than a rule. One way to theorize the specificities of the Moroccan case is to compare it with the situation in other countries. The mechanisms of exclusion and inclusion that make migrants' political incorporation possible make the Moroccan case distinct from those in other countries in the region. As will be explained extensively in the next chapter, the case of Turkey exhibits a different mechanism linking the production of migrant illegality and migrants' experiences of incorporation.

4   Turkey

Depoliticized illegality and a quest for legitimacy

Alima (34) was born in Eritrea and came to Turkey in 2008, after working in Saudi Arabia for three years and in Syria for two years. 'I could send money to my mother but in those places, there is no freedom for migrant workers. I decided to go to other places [...]. The entry to Turkey was difficult, we walked from Syria to Antakya; it was a long, tough walk.' Alima paid her smugglers around 500 dollars to cross the border and another 500 for the bus ride from Antakya to Istanbul. She was held by the smugglers in a safe house, a 'shock house' as they call it. As she was unsure about going to Europe, Alima refused to pay to cross to the EU and managed to get rid of the smugglers, fighting with them until they eventually let her go. She first found a job in an African restaurant in Kumkapı (a neighbourhood in the tourist area of Istanbul, known for large numbers of migrants). 'This is where I met the father of my baby. He is from Nigeria, we moved in together and lived in Kumkapı, Avcılar, Fatih, in different places.' The first time I met Alima, the father of her baby, whom she refers to as her husband, was arrested for selling drugs while she was pregnant. 'He says that he was just next to his friends and was not doing anything wrong but nobody listened to him.' Alima had her baby in 2010 and expected that she and the baby would become Turkish citizens. She later realized this was not the case for her or her child. Before the birth, she was advised to apply to UNHCR. 'I applied for birth, I gave birth in a hospital and the church paid for it.' Alima was able to rent a house with the help of a faith-based charity organization in the Tarlabaşı neighbourhood, right next to Taksim Square, where groups of internal and international migrants reside. 'Then, they sent me to my satellite city, to Antakya. I stayed there 3 months and came back to Istanbul. I was called a couple of times by the police in Antakya.' Alima was able to leave the city by convincing the police that she had to visit her husband in prison or by saying that her baby was sick. She felt guilty for lying and prayed to God that her baby did not really get sick. She shared the house in the Tarlabaşı area with other Nigerians, the 'brothers' of her husband. Alima could not work after the birth of her child and mostly relied on humanitarian aid from the church and other humanitarian NGOs. She later found a job as a translator for an NGO. She was able to leave her child in church-run daycare while she worked. She travelled many times between Istanbul and Antakya while applying for asylum and finally got refugee status. Currently, Alima is in a satellite city, waiting to be resettled in a third country. She still keeps her room in Istanbul and sublets it while she is away.

This chapter traces the interlinked processes of the production of migrant illegality and migrants' experiences of incorporation in Turkey. By exploring migrants' experiences of deportability, participation in economic life, and access to fundamental rights, I underpin processes of low levels of politicization of immigration-related issues in Turkey. As hinted in Chapter 2, I suggest that the low level of politicization, in the general sense that immigration is not high on the public policy agenda, has not only characterized irregular migration governance, but also migrant incorporation in Turkey.

Research has already revealed that Istanbul has become an economic hub for migrants of diverse nationalities and legal status who are looking for economic opportunities. As an outcome of the production of migrant illegality, the irregular migrant labour force has become part of the labour market. The contrast between Harun's relatively smooth absorption into existing textile work in Zeytinburnu, narrated in the Introduction chapter, and Alima's partial access to income generating activities exemplifies my proposition that the labour market is selective and cannot account for the incorporation experience of all migrants of irregular status.

Despite the rigidity of legal and policy categories discussed in Chapter 2, Alima's and Harun's stories do not fit the typical trajectories of a refugee running from conflict, an economic migrant coming to Turkey to work, or of an irregular migrant with the intention of crossing to Europe. At the same time, their stories contain elements from different migrant categories similar to those with irregular status. Indeed, most research and reports on Turkey and elsewhere acknowledge the convergence between categories such as irregular labour migrants, transit migrants, and asylum seekers (İçduygu and Bayraktar 2012; Danış, Taraghi, and Pérouse 2009: 465-6; Biner 2014). However, only a few studies empirically show how drifting from one status to another impacts migrants' access to rights (Coutin 2003), how the strict and legally institutionalized separation between asylum seekers and irregular migrants manifests itself on the ground, or the effects of the politicization of issues pertaining to the rights of irregular migrants.

Observations indicate low levels of migrant political activism and pro-migrant rights movement in Turkey (Parla 2011; Şenses 2012; Ozcurumez and Yetkin 2014; SRHRM 2013: 17). This is surprising given the similar experiences of being stranded due to the difficulties of crossing to Europe

(Yükseker and Brewer 2011) and experiences of marginalization in social and economic life (Danış, Taraghi, and Pérouse 2009; Dedeoğlu and Gökmen 2010). Meanwhile, no research has explored the link between particular manifestations of migrant illegality, migrant incorporation, and (the lack of) communal strategies to access rights and legal status.

How do migrants in irregular situations in Turkey experience their lack of legal status? Why do they not display a similar level of mobilization for rights and legal status as their counterparts in Morocco? Mirroring the structure of Chapter 3, this chapter has four sections. The first explains how migrants experience deportability, as they attempt to enter and exit the country and/or settle in urban areas. The section elaborates on the experiences of deportability and on the disciplining effect of arbitrary practices of the security forces. The second section explains how the mere possibility of deportation profoundly affects migrants' social and economic incorporation. It shows how the precarious structure of the labour market impacts individual experiences and discusses the layers and limits of labour market incorporation. Here, I reflect on the connection between labour market conditions and possible migrant activism. The third section focuses on the difficulty that migrants with irregular legal status have in terms of accessing basic public services, such as healthcare and education where I place a particular emphasis on interconnecting asylum and irregular migration regimes. The role played by civil society is discussed in this section, is elaborated further in the fourth section. Section four also examines how civil society has limited itself to potential asylum seekers, rather than embracing a more radical discourse on the rights of migrants regardless of legal status. This focus has potentially reinforced the depoliticization of issues pertaining to the rights of irregular migrants. After explaining and reflecting on the absence of mobilization for the rights of irregular migrants, the last section focuses, in particular, on individual strategies used by migrants to gain access to legal status and rights.

4.1   Migrant deportability beyond the EU borders

Harun's gave a precise account of his clandestine entry to Turkey from the Turkish-Iranian border and his arrival in Istanbul, as he wrote about his 'adventures' throughout the journey. As they arrived in the city of Van, after crossing the Turkish-Iranian border without documents, the smuggler showed them the UNHCR office without really explaining much: 'There, it is a foreign thing, they send you to other places and also to Europe.' Following

the smuggler's advice, he went to UNHCR but he did not mention his relatives in Istanbul, where he was staying in Van, or how he crossed the border.
'Then, she [the officer] asked me if I ran away from my family. I said I asked
my family before I left, I did not run away. As she kept insisting on why I
came here, I did not say another lie, and I made it clear that I came to Turkey
to work.' She was suspicious, but she sent them to the police department
for registration and settlement in a satellite city.[1] Harun went to the police
department to register but was never resettled in a satellite city because
his smuggler insisted that they continued the journey without delay. They
were travelling to Istanbul by car with a guide sent by the smuggler, when
the police stopped them. 'We had no authorization to leave the city. We lied
again, saying that we are going to a wedding and would come back to Van
after the wedding.' Luckily, they were not detained, but were taken to the
police station and sent back to Van. The police gave them food and allowed
them to spend the night in the station. 'In the morning, he [the police
officer] woke us up, we went out. He stopped a bus on the way and told
the driver to take us to Van bus station. When we arrived, the station was
empty, it was early in the morning. I told the smuggler to buy us our ticket
to Ankara. He was reluctant at first, we were caught once, he said, next time
they might send you to Afghanistan. The second time was no problem. [...]
We took the bus, nobody asked anything. I was in Ankara next morning, it
was the first day of Eid and I joined the morning prayer with other people
in the bus station. The second day of Eid, I was in Istanbul and I celebrated
it there. This was an adventure.'

Rabia and Halim's encounters with the police have not been as smooth.
Rabia, a widow from Afghanistan in her thirties, came to Turkey with
her younger brother Halim and her 13-year-old daughter. They arrived
by airplane with a valid passport and visa. 'Indeed, we came to Turkey
to stay. Here, the situation is better. You can go to school. I was told by
my sister who lives in the UK that if we go to Turkey, we could live there.
Once we arrived, we went to the police department in Vatan Caddesi to
get a temporary residence permit. We were asked to come back with a
Turkish national as a reference. We did not know where to find this person.

1   As explained in Chapter 2, the Turkish asylum system requires a double registration procedure with UNHCR and the Turkish authorities. The Foreigners Department in the Provincial
Police Department registers asylum seekers and sends them to satellite cities where they are
required to reside. Leaving the satellite cities without authorization from the police is not
possible. Asylum seekers are required to sign in at the police department on regular basis to
prove that they are abiding by the rules.

Then, our troubles have started.' Their interview took place three months after their arrival. Within this period, the family's passports were stolen, and they were caught undocumented and detained by the police. They were maltreated by the police, first in a police station, then in Kumkapı Removal Centre in Istanbul. In Kumkapı, they applied for asylum: 'We did not ask for asylum application,' explains Halim, 'they did it themselves, so that they could let us go.' They were happy to be out of the removal centre. However, the police department in charge of asylum assigned them two different satellite cities although they were relatives. 'We were told that it would take at least five months to change our residence to another city. Five months is too long, and I did not want to be alone during this time,' said Rabia, explaining why they decided to return. Thanks to one of the translators, they were informed about the voluntary return programme by IOM. The programme would fund their return, and they were eligible for a non-refundable, one-time payment to make a fresh start in Afghanistan.

These narratives show how migrants who depend on particular ways of entering and staying in Turkey are rendered deportable. As Harun's and Alima's stories imply, the eastern and south-eastern borders of Turkey have been subject to fewer controls. They have been more permeable for potential asylum seekers and economic migrants. Hence, the majority of migrants without necessary documents enter through these borders with the help of smugglers, while a smaller group of illegal entrants, are known to enter via the sea border, crossing the Mediterranean. As in the case of Morocco, those with financial means, connections, and the aim of crossing to Europe move directly to the western borders and try to cross into Greece. Others with no such intention and/or resources to cross, or whose attempts have been unsuccessful stay in bigger cities. They join groups of semi-settled migrants working in the informal market and/or apply for asylum.

Those with valid documents to enter the country mostly come to Turkey on tourist visas and overstay. Others, especially from neighbouring countries, have kept their status legal by moving back and forth between Turkey and their countries. Most migrants, however, overstay their visas and cannot return to their countries because they know that they cannot avoid the stamp on their passport, banning them from legally re-entering Turkey in the near future. Consequently, irregularity becomes a permanent condition for many migrants who overstay their visa.

Regulations preceding the LFIP, as explained in Chapter 2, have illegalized the multiple exits and entries that enabled circular mobility between Turkey and neighbouring sending countries. This change was coupled with an amnesty in summer 2012 that permitted one-time regularization for overstayers. Thus, a small minority of irregular migrant workers could legalize their status by applying for residence and work permits. Others have been pushed into illegality, as they would no longer be able to move back and forth between Turkey and their countries of origin. For instance, Victor, a 30-year-old father of one from Georgia was able to make use of the exceptional non-renewable residence permit scheme introduced in the summer of 2012 after the change to laws on multiple entry: 'The first time, I came with a three-month visa, I renew it by re-entering. Then, I took a resident permit for six months. Then, I was clandestine.'

The functioning of the asylum system is another factor pushing potential and actual asylum seekers into illegality. The regulation requiring asylum seekers and recognized refugees to reside in a remote city (officially labelled 'the satellite city') is intended to create a system of control over asylum seekers and serves to block their mobility to big cities and the EU border. In this sense, the asylum system is designed to filter asylum seekers from irregular migrants (Biner 2014). Meanwhile, as the number of applicants increases and resettlement quotas shrink, the waiting period for resettlement is getting longer. Moreover, asylum seekers who are required to reside in satellite cities face difficulties, especially in finding work to support themselves. NGO representatives underscore that asylum is not appealing for migrants who have not decided whether to settle and work in Turkey or go to Greece. For some groups, asylum becomes another way for migrants to collect documents that enable them to negotiate their deportability. 'We receive applications from Pakistani and Bangladeshi. They apply and never show up again. They just get the paper in case the police stop them. People apply as a last resort, when they are caught, get sick, etc.' (author interview with HRDF, Istanbul, August 2013). While asylum only provides partial enjoyment of fundamental rights (as explained in Section 4.3), most satellite cities do not provide employment opportunities. As opposed to the intent of the legislation, employment opportunities that pull asylum seekers to big cities result in a blurred distinction between irregular labour migrants in bigger cities and asylum seekers, who are legally required to reside in their designated provinces. In other words, the geographical limitation and malfunctioning of the asylum and resettlement mechanism forces asylum seekers to either attempt the journey to Europe through smugglers or breach asylum regulations by moving to bigger cities.

Apprehensions, detentions, and deportations form the main tools of migration control in Turkey (Grange and Flynn 2014). An overwhelming majority of migrants are apprehended by Turkish security forces at the EU borders, along the Aegean coast, and in the Thrace region – the land border between Turkey, Bulgaria, and Greece (İçduygu and Bayraktar Aksel 2012: 24-5). Statistics on migrants' deportation by security forces reveal that the majority of deportees are apprehended while entering or exiting the country without proper documentation (Toksöz, Erdoğdu, and Kaşka 2012: 47). The analysis should acknowledge that the data on irregular migration is far from complete or reliable (SRHRM 2013: 17). Moreover, the geographical and porous character of the borders in eastern Turkey challenge nation-state controls over these boundaries. Nevertheless, one can deduce from available statistics that the Turkish state has been preoccupied with controlling irregular exits, which is also the EU priority, in the context of the international production of migrant illegality, as explained in Chapter 2.

Looking at the 'geographies of deportability' (Garcés-Mascareñas 2012: 210) and 'geographies of detention', one can suggest that there are differences in the experiences of deportability of migrants attempting to cross EU borders and those who are semi-settled in urban areas. Istanbul, where migrant interviews for this research took place, has been identified as a major hub for migrants who stay and work in Turkey without proper documents (Danış, Taraghi, and Pérouse 2009; Suter 2012). Yet, deportees who were apprehended and detained in Istanbul constitute a small portion of all deportees throughout Turkey (Toksöz, Erdoğdu, and Kaşka 2012: 81). According to data collected by Toksöz, Erdoğdu, and Kaşka (Ibid.) for an IOM report on irregular labour migration to Turkey, out of 44,433 deportees in 2001, 10,795 were deported from Istanbul; and out of 26,889 deportees in 2011, 8,592 were deported from Istanbul. Also note that the number of deportations of 'undocumented' workers has grown over the years, but has not exceeded ten per cent of total deportees. For Toksöz, Erdoğdu, and Kaşka (2012: 48), the relatively low percentage of undocumented workers deported is due to lax inspections. Tolerance of irregular migrants, especially those who are working informally, is a widely acknowledged aspect of detention practices in Istanbul. The available statistics, albeit inadequate, give a general impression, but do not offer conclusive information. Their interpretations should benefit from the insight gained from migrants' own perceptions of illegality.

Experiences of deportability: Between tolerance and arbitrariness

A survey conducted with over 1,000 foreigners with different legal status, residing in different cities in Turkey, indicated that 64 per cent of respondents agreed that the police in Turkey are tolerant towards immigrants. Surprisingly, migrants who did not enter the country through legal means were the group who agreed most with this statement (74 per cent) (İçduygu et al. 2014: 80). Victor, a 30-year-old young man from Georgia articulates his perception of police tolerance thus:

> Police ask for passports and identity all the time. Everybody knows that I am working *kaçak* ('unregistered', 'clandestine'), nobody touched me. I used to work in a car wash. There I was also washing policemen's cars; the police came and naturally we talk, have tea. They all know that I am here to work, I do not do anything bad, I do not mess around.

This sub-section elaborates on the link between migrants' perceptions of being tolerated and on how this perception impacts their experiences of illegality and of incorporation. The narratives of being tolerated are enmeshed with those of being subject to arbitrary practices. What I call 'arbitrary toleration' by security forces refers to a myriad of practices, ranging from turning a blind eye to the presence of irregular migrants to opportunistic crimes, such as occasional controls that lead to harassment or bribery, if not detention (also documented by Suter 2012; Yükseker and Brewer 2011). After arriving in the city, the post-entry period is characterized by the possibility of detention, but also by widespread tolerance, especially for groups that are relatively less associated with transit migration. Even Afghan nationals, who are overrepresented in deportation statistics and considered a group that is likely to be subject to long detention (Toksöz, Erdoğdu, and Kaşka 2012: 48; SRHRM 2013: 12), have expressed that they feel they are being tolerated, as long as they stay away from the borders and from crime-related incidents. A police officer once stopped Malik, a 20-year-old man from Afghanistan, to ask for his papers: 'He did not ask anything more, when I said I am from Afghanistan,' he explained to me with a sense of security. In a similar vein, although Harun, another young man, also from Afghanistan, defines himself and their way of life as *kaçak,* he also expresses that being *kaçak* may not necessarily affect one's relations with the security forces:

> I have been in Turkey now for three years. We are living here, we are living *kaçak*, we are in comfort here, we are working. Thank God, it is like our country. If the police were to see me, they would let me go. [...] Indeed, once they caught me, but let me go. They were investigating drug use, and there were men doing illegal business. They let me go.

Despite the harsh situation at the borders, the experience of deportability in the urban setting in Istanbul is characterized less by raids and more by arbitrary law enforcement. According to CSOs operating in Istanbul, raids are only used occasionally in criminal situations or when the police have been notified (see also Toksöz, Erdoğdu, and Kaşka 2012: 82). The implementation of detention practices in Istanbul continues to be unpredictable. As explained at the beginning of the section, the police detained Rabia, her brother, and her daughter because they were not able to present any legal papers: 'We were arrested as we were walking by the seaside. We said we were Afghans, he asked for papers. We tried to explain that our papers were stolen. They never listened, we were put in the police car.'

Interviews with security forces on the practices of control confirm the possibility of being subject to random controls, rather than systematic inspection. The security forces themselves also affirm the situation of relative tolerance towards certain groups, but especially to certain types of mobility. An officer from the Turkish National Office explained that police raids target *şok evleri* (safe houses) in Basmane, a district in Izmir on the Aegean coast, and in the suburbs of Istanbul where smugglers keep migrants before they organize their crossings out of the country. Raids of this type directly target those who might be subject to transit migration. Alleged economic migrants are also subject to police controls but to a lesser extent. The officer interviewed highlighted the impossibility of entering houses: 'Inspections are done in public spaces, in entertainment places. The controls are done through profiling, there are no controls for every one we meet in the street.' Referring to African migrants in Istanbul and Ankara in particular, he added that migrants of irregular status are depicted as a problematic group in terms of security, 'prone to criminal activities' especially when 'they cannot earn money' (author interview with an officer from the National Security, December 2012, Ankara).

The random character of controls by the security forces is justified by a lack of institutional capacity, but also a lack of interest in detaining migrants, especially in urban areas. 'The police does not make raids in Istanbul, what would they do with those apprehended?', an NGO representative stated, expressing the general response from of NGOs when asked about police behaviour. This conviction is underpinned by the fact that the official capacity of removal centres has been much lower than the number of people apprehended, although it is known that the actual detention capacity, including police stations used as de facto spaces of detention, is higher than reported (Grange and Flynn 2014: 19).[2] While academics have

---

2   DGMM indicates that the detention capacity will reach from 6600 to 13,660 by the end of 2016. Retrieved 21.11.2016 from http://www.goc.gov.tr/icerik3/removal-centers_915_1024_10105

calculated the latter number to be much higher than the figures given by the European Commission (EC), it still indicates a low capacity when compared to the annual number of detentions. As detention conditions have been a major focus of international and domestic critiques, one can fairly argue that the police have less interest in detaining mass numbers of migrants. Officials also referred to migrants' contributions to the economy, and the fact that Turkey is becoming a country of immigration, as grounds for tolerating migrants' economic presence, even those lacking papers. It should be noted that this justification of tolerance based on the economic contribution of migrants was also used in the aftermath of the Syrian crisis (Özden 2013; Mazlumder 2014).

Being tolerated does not necessarily mean being free from detention and deportation practices enforced by the police. Indeed, along with *kaçak*, another word commonly used among migrants from different migration trajectories is 'deport'. Most migrants use the word regardless of their knowledge of English. Indeed, the police have been given wide discretionary powers to interpret public order, which means that whether police intervention results in detention or deportation depends on the enforcer on the ground. However, being a foreigner in and of itself may provide sufficient grounds for detention if a legal migrant has been in the vicinity of an incident:[3]

> Let's say two people – one citizen, one non-citizen – are taken into custody. They are released afterwards based on a decision by the prosecutor. Only because he is a foreigner, the non-citizen is not released but sent to the foreigners' police, simply because he was involved in an incident with the police [...]. I also encountered cases where the victim complained to the police and then faced detention followed by deportation. In this sense, the prerogative of the police is limitless. The fact that it is a foreigner [regardless of legal status] provides enough grounds to deport the person. (Author interview with an advocacy NGO, Istanbul, November 2013)

Being tolerated and being subject to arbitrary practices and abuse go hand in hand. In the context of the wide discretionary powers of the police and the absence of judicial mechanisms, it is challenging for irregular migrants

---

3   One striking example of arbitrary detentions happened during the Gezi Park protests in June 2013. See Student Detained during Gezi Protests Appeals Deportation Decision. Retrieved 29.03.2015 from http://www.hurriyetdailynews.com/french-student-detained-during-gezi-protests-appeals-deportation-decision.aspx?pageID=238&nID=49639&NewsCatID=341

to contest violations of their rights. Dilbar is a 33-year-old woman from Uzbekistan, working in childcare in an affluent residential neighbourhood in Istanbul. Like other women from the region who come to work in Turkey, she came on a tourist visa and found her first job through an informal employment agency. Even after her passport was stolen because she had lent it to a friend to whom she was indebted, she was able to change jobs. Another Uzbek woman from her village recommended that her new employer accepted her without a passport because she had good references: 'It is so hard to find a job without a passport. They trusted me thanks to sister Shara.' She explains that because she is a woman, she refrains from being outside after dark and mentioned that she was actually stopped by the police. She used to go to out on her days off, during the weekend, and also spent nights there with other Uzbek women who worked for the same agency. On her first payday, she went to pay the informal agency the fee for finding her a job. On her return, she was stopped by the police and was put in a police car:

> I told him I was working, I was from Uzbekistan. He asked everything, and I did not answer much, only short answers. 'Why you do not talk?' 'Don't you have a tongue?' I was so scared, I could only answer, 'I do.' He asked me if I had money. I said I did not, at first, and I refused to give him my purse. Then, I said I had little money. I told him how much money I had. He asked for half of it. So, I gave him 200 dollars and kept the other 200 and returned home.'

The threat of detention and deportation was enough for Dilbar to give the officer her whole wage. Interestingly, the police 'only' asked for half of her wage and assured her that she would not have to pay the next time. This instance exemplifies how tolerance implies inherent arbitrariness by enforcers and the possibility of abuse. Oral statements such as the promise that she will no longer be asked for money indicate how practices of 'arbitrary toleration' towards irregular migrants render them docile.

Meanwhile, migrants themselves are not without counter-strategies. The awareness of arbitrary detentions and abuse by security forces has led migrants to make conscious decisions about how to behave in public and what kind of papers to collect in order to show that they are not involved in criminal activities, including irregular border crossings. As a result, migrants feel their deportability on daily basis, and they make conscious efforts to negotiate the discretionary power of the police. These efforts may include the collection of certain papers – a passport with a valid entry, asylum applications, other identity documents – indirectly proving their length of stay in Turkey and that they are not prone to criminal activities or transit migration. Most Afghan nationals spoken with, who enter the country without legal documents, explained that, upon their arrival, they felt the need to go to the Afghanistan consulate in Ankara and apply for an identity card showing that they are from Afghanistan. This passport, commonly called a 'white passport', written mainly in Farsi, is not useful for travelling or obtaining a residence permit in Turkey, but it is sufficient for showing the police. The date of issue of the document indirectly proves the time spent in Turkey, hence, the person is not a likely candidate for irregular border crossing into the EU. As explained by Malik, the possession of certain papers, albeit not the *right papers,* may be tolerated by the security forces: 'Yes, the police stopped me once and asked for my identity, 'Unfortunately, I do not have one,' I said. He asked for my residence permit. 'Unfortunately not.' Then, he asked me what I had and I showed my passport. He looked at it and let me go.' One of Malik's flatmates, Ahmed, the eldest son in his family, came to Istanbul to work in 2012 at the age of 19, after many years of being a refugee in Pakistan and later in Iran, where he worked for nearly two years. He explains that the police have become more tolerant, compared to Iran, especially within the neighbourhood that is known for the presence of irregular migrants:

> The police may stop you and ask for your identity. We, Afghans, can show our passports to the police, and it is not much of a problem [...]. When I came here, I saw my friends had gotten their passport. I also went to Ankara to get one for myself [...]. So far, the police have never stopped me, but my friends were stopped. It is because I work in Zeytinburnu and only leave the area once in a while, during Eid or something like that.

By collecting certain papers and selectively using the urban space, migrants consciously redraw lines between legality and illegality, and situate themselves in this illegal but licit sphere (Kalir 2012: 27). On one of my early visits to a house where single Afghan men lived and worked, in a district of Istanbul known for textile and leather workshops, I noticed a tense atmosphere and a reluctance of people to speak with me, something I had not previously encountered. Then, someone explained that the police had arrested their friends who were living with a smuggler. There had been a police raid in their friends' neighbourhood, and the police apprehended them. Three of them were released immediately (reportedly by saying that they are 15 years old): 'The police will release them, it will not do much.'

Mahmut, a 29-year-old man, the eldest in the house, who usually responds to my questions on behalf of everyone, explained to me that the police were searching for one particular person, an alleged smuggler, related to a recent boat sinking that killed 12 people. His explanation was followed by a cautious remark: 'Obviously, we know those kind of people, but we do not let them hang out with us. Being around us, they constitute a problem for us. As we are working here, we do not let those people among us.'

Even among African men who are stigmatized as prone to criminal activities, I have noticed a similar sense of confidence emerging in a relatively short time after arriving in Istanbul. The general conviction, expressed, for instance, by Alex from Nigeria, is that 'the police would not touch you if you are not doing anything illegal' or 'migrants are not harassed in Turkey'. These convictions reveal that even the groups that are most stigmatized in the media and by police practices feel at ease (see also Suter 2012: 122-126). Apparently being tolerated, however, is coupled with occasional abuses, checks, and the possibility of detention. Being *kaçak*, hence deportable, refers to migrants' subordinate position in social and economic life. In the absence of channels through which to claim their rights as human beings and/or workers, they are left at the mercy of the police and their employers. Market participation is one way to show the police, and implicitly the locals, one's docility, and hence one's legitimate presence, despite being *kaçak*. Docility describes more than an image created for the security forces and locals; it also characterizes migrants' experiences of settlement and labour market participation and their quest for rights and legal status.

4.2 Illegality in (semi-)settlement: Incorporation into informality

The existence of a widespread informal economy and community networks facilitates processes of settlement and economic participation. The structure of the labour market, which is informal and requires cheap labour, has enabled the economic participation of irregular migrants. In dialogue with the findings of research on irregular migrants' economic participation in Istanbul (Toksöz, Erdoğdu, and Kaşka 2012), I suggest that migrants' labour market participation contributes to the construction of docile subjects; additionally, I show that this is a selective process. Namely, the labour market is more likely to accommodate a young, flexible labour force that can survive tough and precarious working conditions. Surprisingly, neither migrants' experiences of 'market violence', commonly expressed in interviews, nor their exclusion from economic activities translates into political contestation.

Settling into informality

The availability of housing and labour market opportunities and the presence of community networks have enabled the concentration of migrants in certain areas. These neighbourhoods, such as Kumkapı, Kurtuluş, Dolapdere, Tarlabaşı, and Zeytinburnu, are mostly situated at 'the periphery of the centre' (Danış, Taraghi, and Pérouse 2009: 469). They have historically shown ethnic and/or religious diversity. Housing is available through mechanisms of the informal economy and ethnic kinship ties with those already settled. Most of these areas are close to districts where migrants find work opportunities. Neighbourhoods such as Kurtuluş and Kumkapı are close to small-scale manufacturing ateliers and cargo businesses, widely used for textile trade. Zeytinburnu is largely inhabited by communities of migrants of Afghan origin who acquired citizenship after arriving in the early 1980s, as well as more recently arrived migrants and asylum seekers from Afghanistan, comprising a heterogeneous group in terms of legal status, ethnic, and linguistic background. Citizens of Afghan origin and the newcomers in the post-2000 period work in small-scale leather and textile manufacturing, which was already common in the area (see Danış, Taraghi, and Pérouse 2009).[4]

Pure market forces and lax regulations define migrants' inclusion into the housing market. Announcements – widespread in the streets of neighbourhoods such as Kumkapı, Aksaray, and Kurtuluş – advertise housing that is available for foreigners on shop windows, conveying familiarity with foreign newcomers. Housing is a form of economic revenue for property owners and real estate agents who may ask immigrants to pay higher rents. The increasing of rents with the arrival of more immigrants without legal status in certain areas reveals one way in which migrants contribute to the local economy (Biehl 2015). Eric, a 34-year-old man from Cameroon, came to Turkey with a valid visa and open return ticket to look for possibilities, including crossing to the EU, securing legal papers to travel to the USA, and/or trading between his country and Turkey. His 'brother', also from Cameroon, welcomed him at the airport and took him to Kumkapı: 'I am staying at a friend's house. There, we are four. We pay 100 dollars per person

---

4   In 1982, a Turcoman community from Afghanistan was invited by the authorities to settle in Turkey. In later years, they were joined by newcomers as a result of marriages and family reunification (Danış, Taraghi, and Pérouse, 2009: 543-4).

to cover rent and bills. The landlord is fine with us staying there. The land-lord only cares about the money.' For Harun from Afghanistan, it was easier to rent a house with his parents and siblings. However, he was concerned that the arrival of Syrians and ongoing gentrification in the Zeytinburnu area would raise rents even further. 'We took an apartment eight months ago. It is easier to find an apartment if you are a family. However, there has been a lot of new arrivals, Syrians recently, and the rent prices went up. I am not saying anything against them, they should be able to come, like we did. Plus, a part of Zeytinburnu will be evacuated so the rent will go higher and higher.'[5] The ongoing urban transformation of Istanbul has limited housing possibilities for the lower classes, including immigrant groups.

Most migrants who arrive in Istanbul find housing through relatives or ethnic kin who speak their language, whom they may or may not know before arriving. In these cases of semi-informality, older and more estab-lished migrants can help their newly arrived kin through their legal status. Landlords may ask for legal papers for rental contracts. In this case, former migrants who have (already) acquired legal papers may help newcomers to get contracts, at times in exchange for money. Ahmed, for instance, originally from Afghanistan, got help from kin from his village, who left for Turkey back in the 1980s and became citizens: 'I had relatives living here. I called them and let them know about my arrival. We were three people on arrival. They rented a flat for us, we found the house, we found work, now we are all working.'

It is noted that some new arrivals without connections, and those with the intention of crossing into the EU, may live with their smuggler. They are usually overcharged for accommodation and, at times, may be pulled into a debt bondage. Dilbar, a domestic worker from Uzbekistan, was not able to find employment as a domestic when she arrived. She stayed in an informal labour agency in Kumkapı and paid a daily rate for the bed she was provided with while looking for a job. Zerrin (34) and her children moved from Afghanistan, first to Iran and then to Turkey, with the help of smug-glers. She hopes to go to Europe legally to reunite with her husband, who received refugee status in the Netherlands after spending years in different EU countries. Upon arrival, Zerrin did not know anyone in Istanbul, but their smugglers established connections with her husband's 'friends' and found them a flat to rent in Zeytinburnu. Zerrin, unlike other migrants

---

5   See also, *'Somali Sokak' 'Suriye Sokak' oldu* ('"Somali Street" has become "Syrian Street"'), for a media account on the impact of arrival of first Africans and recently of Syrian migrants on rent prices in Kumkapı area. 10.08.2014, *Hürriyet.* Retrieved 22.03.2015 from http://www. hurriyet.com.tr/ekonomi/26939576.asp

with initial connections, had to pay for this service. Over a year later, Zerrin and the children are working in textile ateliers in the neighbourhood and are still paying their debt to the people who initially found them a house.

It is also common to sub-let rooms in a house once an initial contract has been agreed. Newcomers can settle into overcrowded flats or rooms that are inhabited by migrants who arrived previously. After signing an initial contract, the house that Ahmed's relatives rented was inhabited by a group of single men from Afghanistan. Another Afghan family that was smuggled from Afghanistan settled in Zerrin's second room. Accepting newcomers is a way of reducing the cost of rent and raising revenues for more established migrants. Alima, a recognized refugee and mother of a three-year-old, from Eritrea, whose story is briefly introduced at the beginning of the chapter, has resided in Istanbul for years, rather than in her designated satellite city. As mentioned, she initially rented her two-room flat with the help of a church-related charity. She dealt with the property owner and newcomers in the house, who were mostly Nigerian men. When I first met Alima in 2011, she was paying 200 TRY (New Turkish Lira) per month (around 90 Euros in June 2012), half of the total rent. During one visit to Alima's house, in 2013, on a day the rent was due, Alima, a bit confused, and intensively calculating the correct amounts to collect from each resident. When I helped her count the money, I realized that she was paying less than before and that every tenant was contributing different amounts. When we were alone, I asked her about the discrepancies in rental payments. 'It depends on the work' she said, 'some people have better jobs.' We later discussed the fact that it also depended on when the tenants had arrived; newcomers were expected to contribute a bit more compared to others.

My methodology and conceptual framework have been limited to re-vealing relations with locals and other migrant groups in neighbourhoods significantly affected by immigration, commonly characterized by 'illegal' activities such as human smuggling, drug dealing, prostitution, etc. According to a survey conducted with African migrants in Istanbul in the early 2000s, 29.5 per cent of respondents indicated ill-treatment by strangers as their most common problem in Istanbul (Brewer and Yükseker 2009: 702). As revealed by ethnographic research, migrants have been subject to stigmatization and opportunistic crimes and are uneasy about being outside, especially after dark (Suter 2012: 131-2). Black migrants, especially Nigerians, have been identified as drug dealers (Suter 2012: 117). Women from post-Soviet countries have long been stigmatized as 'Natasha', the name commonly attributed to those coming to Turkey as sex workers. Recently, African women have also been seen as sex workers. They are subject to sexual harassment on a daily basis, in the streets and in the workplace. Blessing, a Nigerian woman in her late 30s, explained that she was reluctant to commute to work in Avcılar (a district towards the western end of the city) and return home at night. 'There are a lot of *alibabas*,' she said, referring to neighbourhood gangs stopping her and asking for money. Men, from African countries in particular, articulated discomfort with approaching 'white women' and interacting with locals outside of their work relationships. The threat of neighbourhood violence exists, but migrants of irregular status have few channels to articulate their suffering. Non-state actors confirm that women in particular are unprotected and cannot go to the police when they are sexually abused; this is especially true for Africans. According to reports, cases of racist attacks have not been widespread, but they occur and are rarely covered in the media.[6] Notably, the media have started to cover discussions of xenophobia against immigrants and refugees in Turkey due to increasing tensions between Syrians and locals (Şimşek 2015). Section 4.4 discusses a few instances of the recent politicization of racist violence. No comprehensive research focuses on xenophobic violence targeting immigrants in the Turkish context, so it is hard to determine whether migrants are unwilling to express their experience of violence, or if cases of violence are indeed sporadic. It was surprising that violence towards migrants was not at the centre of migrants' narratives of neighbourhood life, while migrants have more openly expressed their suffering due to labour market conditions, as discussed in the next sub-section.

'We arrived, slept, and the next day we started working'

Earlier research on irregular migrants' labour market participation in Turkey has extensively focused on domestic work and gendered aspects of irregular labour migration (Akalin 2007; Kaşka 2009; Keough, 2006). There has been less research on migrants' roles in the precarious workforce in the urban informal economy (Arı 2007). Thus, the focus has been placed on the migrant labour niches, in sectors such as care and tourism (Gökmen 2011; Toksöz and Ulutaş 2012), rather than migrants of irregular legal status working among the urban poor. For groups such as domestic workers, economic participation results from a concrete demand by employers, as is well documented in the

---

6   Four migrants from Liberia were shot and injured as a result of a racist attack in Istanbul. The media covering the issue drew attention to the sporadic nature of these events. See for instance, Liberians injured in a racist attack. 11.04.2014, Retrieved 22.03.2015, from http://www. dailysabah.com/nation/2014/04/11/liberians-injured-in-a-racist-attack

literature. It is surprising that other groups of migrants who are categorized as being in transit are equally present in the labour market, regardless of their initial intention to work in Turkey or to cross to Europe (Toksöz, Erdoğdu, and Kaşka 2012: 20). I suggest that for some migrant groups who came to Turkey for purposes other than work, such as asylum or crossing to Europe, participation in the labour market is an unintentional result of the interaction between the production of migrant illegality within the international context and the domestic structure of the labour market. In his study of the male Pakistani and Afghani labour force in London, Ahmad (2008: 872) makes the point that 'the structure of the labour market that absorbs new migrants does not always make clear distinctions between the so-called "legal" and "illegal" migrants.' This observation also applies to the case of Istanbul, where legal status is not the most important factor determining labour force infiltration. Migrants with temporary residence permits (but without work permits), overstayers, asylum applicants, and undocumented migrants smuggled into Turkey have similar experiences in the labour market.

As an unintended consequence of external dimensions of EU migration policies, even migrants who are most likely to go to Europe participate in the labour market during their unknown waiting period in urban centres in Turkey. In general, migrants' labour market behaviour makes it difficult to determine whether they intend to cross to the EU or to stay. Note that in some cases, houses that are unfurnished, not even with a sofa, generally indicate that the people or the family have either just arrived, trying to decide where to settle, or they are more interested in investing in their journey than in settling. Said, for instance, is a young man from Afghanistan who came to Turkey with a valid visa about a year before I met him in one of the houses inhabited by single Afghan men. A 16-year-old teenager, he was always shy and reluctant to answer questions. One Sunday afternoon, I visited their house. We were discussing the difficulties of life in Istanbul, and I asked if they knew anybody who planned to go to Europe, and everybody looked at Said. 'His father has decided that Said should go to Europe. It is good for him,' explained Mahmut, on behalf of Said. Like all the other single men in the house, Said had been working while waiting for opportunities and financial aid from his father to further his clandestine journey to Europe. As mentioned above, Zerrin, a mother to two from Afghanistan, had started working in textiles, while trying to understand how the asylum system in Turkey functioned, with the intention of reuniting with the father of her children in Europe. Among the families in Zeytinburnu, there are some whose children or younger family members worked in textiles while they were trying to renegotiate a deal with smugglers to take them to Europe.

Migrants, including those on their way to Europe, have become part of this labour market as an outcome of EU migration controls, as also implied in other research (Yükseker and Brewer 2011; Danış, Taraghi, and Pérouse 2009; Suter 2012). When employers know that migrants intend to cross to the EU, they may use the situation to their advantage. They may justify low wages and hard working conditions, saying that they are actually helping those stranded in Istanbul while they are in transit. In one of my initial visits to Kumkapı in Spring 2012, I encountered an African man from the Ivory Coast in front of a cargo company in the area. It was easy to initiate a dialogue, as he was happy that he could speak French with me. He immediately expressed his intention to go to Greece and then France. He had already been in Istanbul for eight months and had been working for the cargo firm for a couple of months. He did not know when he would be able to leave. In the middle of the conversation, the shop owners, who had been in the cargo business for two years, approached us. One of them explained that they did not have anyone working with them and said, 'then came this *kara çocuk* ('black boy').' He continued, 'he needed a job and we gave him a job. He is not doing much anyway. We are helping him because he needs money.' It is striking how the owner presented the situation as if the 'black boy' was working there due to the benevolence of the employers, most probably for relatively low wages. Precarious employment is a general characteristic of the labour market, and migrant illegality is embedded in the labour market, as further explained in Chapter 2.

In the context of selective labour market permeability, migrants may find jobs through informal mechanisms. Similar to the situation in housing, ethnic kin or informal employment agencies that newcomers have been in contact before arrival or encounter during the settlement process may help them find jobs in the informal sector. People with few connections go from door to door asking for work, a practice usually described by African migrants in the Kurtuluş, Kumkapı areas. Chris came from Nigeria to Istanbul with a tourist visa and the intention to work and study. Like other African men, he wandered around the streets of Merter, another area known for textile shop floors, asking for available work in his limited Turkish: *iş var mı?* ('Do you have work?'). This is actually how Chris found his first job in a textile atelier producing bags. Another migrant, Malik, was lucky that he did not have to look for housing or a job upon arrival in Istanbul, after crossing the eastern border with the help of smugglers.

The brother of my fiancée was here, I had relatives from Afghanistan. We came in, slept and we started working the next day [...]; yes, the very

next day, he told me that this is how it works in Istanbul. I had planned
to rest and take care of myself for one month or two. 'Not here,' he said,
and we kept on working, without any interruption for six months. Then
we quit the job because the money was not enough. We went to another
job in the Grand Bazar, I worked there for another five months. I quit
again. Then, I went into printing.

As Ahmad (2008: 864-5) explains, for some who fit the profile needed in the
labour market, short-term work is more available than regular jobs. Like
Malik's story, it is widely observed that migrants working in leather, textiles,
or construction often have to move from one workplace to another and shift
between sectors. Harun noted that there were fewer opportunities when he
arrived in 2009, due to the aftermath of the 2008 financial crisis. He found
work in leather and textiles, but often had to change jobs because of seasonal
changes in the market and the general flexibility of production in these sectors.

Then, in 2009, the work was scarce in Turkey. I had no jobs for the first two
months. Then, I went to work in leather. It was weekly work. Some weeks,
he [the employer] was giving 130 TRY, some 140 TRY if you do extra hours.
When the leather season was over, I left the job and went into the bag atelier.
I worked there, then the person from the leather place called me back and I
went back there to work. I quit again after some time and went into textile.
The guy called me back several times but I did not return. [...] Leather is
difficult, it takes time to learn how to do leather jackets. In textile, you can
be *usta* ('head person') in six months. Other head persons will help you, the
boss and other friends working there will help you to learn.

During our first interview, Malik was unemployed and looking for a job.
He was searching through ads for work in the textile industry. His age was
an obstacle to employment, and some employers were looking to employ
women: 'For the moment, I am the only one with no work in the house,' he
explained, still hopeful that he would find work soon. Indeed, it did not
take Malik long to find a job in textiles. After around six months, we met
again during Eid. Malik was back in Istanbul to visit his fiancée's family and
celebrate Eid with friends. He and his friend Ahmed (22, from Afghanistan)
both moved from textiles to construction and from Istanbul to Ankara.
Malik later explained that he does not need to pay rent, as he stays on the
construction site, and the daily wage was comparable to textile work. He
also found that the work requires physical power, but was not as tedious as
working with textile machines, which require you to sit for hours without a

break. In private, he told me that he was no longer worried about being *kaçak* because he had begun to use the social security number of his employer's son. Malik considered this to be a form of legalization of his status. I doubted that this arrangement could protect him from the risks involved in his work. Moreover, this case illustrates how employers use fraud to conceal the fact that they employ immigrants without legal status.

Given the availability of work in the lower echelons of the labour market, one can question whether the labour market in Istanbul provides mobility prospects for migrants. The narratives of migrants working in textiles, usually from Afghanistan, indicate that they receive an increase in their wages as they gain skills and experience. Ahmed explained that, in time, he could generate enough income to enable him to send money to his parents in Iran: 'Before, what I gained was not enough for myself, I was earning 600-700 TRY per month. There was the rent and my expenses. Then, I moved to the machine and my wage now is around 1300 TRY. It is very good. I also send money back home.' Note that the minimum net wage in Turkey was around 770 TRY in the first half of 2013. Ahmed 'moved up' relatively fast as he was experienced in textiles; he had worked in a textile atelier after school when his family took refuge in Pakistan. Selma, also from Afghanistan, entered Turkey without legal papers, with her husband and two children, after several years of being a refugee in Iran. She had no prior work experience before coming to Istanbul. She had to work because her husband faced difficulties in finding a regular job in Istanbul. Selma's narrative shows the volatile trajectory of workers in the textile industry, from being an *ortacı* ('a middleperson helping out machine workers') to a machine worker, but also moving from ethnically homogenous to mixed workplaces, which seem to pay better:

> I changed work five times. Before, I did not know the work, I was working with an Afghan, he was giving me 300 TRY. The second workplace was also run by an Afghan, the wage was 600. There, I started to learn the machine. Then, another workplace where the boss was a Turk, everybody was Turkish, I earned 800 TRY per month. Now, I know how to use the machine very well and my wage is 900 TRY per month.

Selma and her family secured residence permits around a year after arriving, through Selma's brother, who had arrived in Turkey as a medical student in the 1990s and gained citizenship. The fact that they were no longer clandestine in terms of residence status did not have an impact on her or her husband's labour market experience. Instead, her economic participation was the result of the existence of a labour market in leather

Figure 4.1   Kumkapı, packing and carrying goods before shipping them overseas

Source: Photo taken by the author

and textiles that was ready to employ newly arrived Afghans, regardless of their method of entry, legal status, or aspirations to cross to the EU. That said, access to legal status facilitates economic activities. Some Afghans in Zeytinburnu with residence permits even managed to open their own small workshops.

Other groups without similar informal reception mechanisms, such as informal agencies or more or less established ethnic economies, may experience marginal forms of economic participation. Among the migrants interviewed, it is observed that those from African countries have a harder time finding employment in what they refer to as *çabuk çabuk* ('quickly/ chop-chop'). These are daily, poorly paying jobs that require the person to work as fast as possible. Chris, like other informants from Nigeria, had an aim to move into trade after securing enough income through his smaller jobs: 'We need money to do cargo business, first you do textile and you can start with cargo after saving some money.' However, Eric, a 34-year-old man from Cameroon, made it clear that the work is sporadic and does not provide a stable income; it is only to survive, and you cannot make enough money to save to cross or send to your family. 'There is little money, you can still manage but you cannot save,' articulates Peter, another migrant from Nigeria who arrived in Turkey by plane with a one-month tourist visa.

Limits of labour market participation

Given the labour market conditions, it is a realm of both exclusion and inclusion. The aspiration to cross to Europe is one reason why some migrants are
less interested in temporary work, which generally does not provide enough
income to finance their journeys (Wissink, Düvell, and Van Eerdewijk 2013:
1099). In other words, some are more interested in arranging the journey as
soon as possible rather than spending long hours in poorly paid, tedious jobs.
For instance, Muzaffar, a 44-year-old man travelling alone, is originally from
Pakistan, but has worked and lived in Dubai for several years. After being
expelled from Dubai for his political activism, he returned to Pakistan and
travelled to Turkey through Iran, with the help of smugglers. Shortly after his
arrival, he tried to cross the border to Greece via the maritime route. When
the attempt failed, he continued to live with the smuggler who had brought
him to Istanbul. At the time of our interview, he was waiting for another
opportunity to cross and simultaneously looking for alternative ways to leave
Turkey, such as resettlement through asylum. Muzaffar was not interested
in what the labour market in Istanbul could offer: 'I came here with Afghan
people, I paid 3,000 US dollars. I also paid for Europe. They promised me it
would be in one month, two months, it's now six months I have been here
[...]. They sucked up all my money. They offered me a job. I said, "I do not
want your job, I did not come here to work"'. Migrants like Muzaffar, who
rely on money received/borrowed from abroad to continue their journeys
may refuse to work in poorly paid, difficult jobs. However, the prospect of
transit migration is not the only reason for exclusion from the labour market.

Despite being open to a young, flexible, healthy workforce, joining the
labour market is a highly racialized, gendered, and sexualized selection
process. Peter, a Nigerian man working in textiles but aspiring to trade goods
between Turkey and Nigeria, wanted to bring his wife; he complained that
there was not much work for a black woman in Istanbul:

> She can work in textiles, she can be the middleperson, do cleaning, but
> Turkish people do not offer many jobs to African women. It is about the
> black skin. If I am a black man, it is more difficult to find a job. If you are
> a bit fairer, then it might be possible. Ethiopians for instance, they are
> fairer, for them it is easy to find *madame work* ('domestic work').

Anecdotally, there have been demands for Ethiopian women in childcare
because they can speak English (author interview with Caritas, Istanbul,
January 2014; also mentioned in Brewer and Yükseker 2009: 699). The East

African women I encountered found opportunities in the care and service sectors, but they had to be extremely cautious because many service jobs entail offering sexual services to employers, intermediaries, or customers.

There is a thin line between engaging in sex work and migrant women using, or being expected to use, their sexuality as a currency in the labour market. Existing research has revealed sexual exchanges between migrant women from post-Soviet countries and their 'business partners' in the context of the suitcase trade (Yükseker 2004; Bloch 2011). However, less has been written about sexualized work by other migrant groups, such as African migrant women, or on sexual exchanges among migrants. As a relatively more experienced migrant, Alima from Eritrea mediates between employers and migrant women (usually from Eastern Africa) looking for jobs. One day, two young women from Ethiopia (one with an asylum application) visited Alima to discuss the opportunity of working in a restaurant run by Nigerian migrants on the outskirts of the city. Alima explained to them that the restaurant is safe in the sense that they will not be asked to offer sexual services. The young women looked sceptical, but agreed to meet the Nigerian man running the place. Alima was supposed to receive 100 TRY (around 35 Euro in 2013) from each for being the intermediary. I later heard from Alima that she did not receive the money because the women refused to work there when they learned that they would be asked to sit with customers. Blessing, a mother of two from Nigeria in her late 30s, decided not to go back to the workshop where she was employed to make jeans because she found the work too difficult: 'You have to stand from 9 am to 11 pm', and her body ached afterwards. She later found work in a kitchen in Avcılar, in the west end of the city, but she was not happy there because it was far, and she had to work until midnight. I asked her what kind of job she wanted: 'I want something that will not tire me out.' She did not like çabuk çabuk because she had a problem with her knees. She wanted to be a salesperson or similar. However, Blessing was not physically suited to this kind of work. Since most saleswomen are also asked to model for overseas customers, S or XS size women are preferred. I later heard that Blessing had quit work and moved in with a Nigerian man who allowed her to stay in exchange for sexual favours. While these live-in arrangements are common among migrant women, their implications for labour market participation are dubious. Kuku, another young woman from Ethiopia was five months pregnant when I met her. She worked as a salesperson in Osmanbey, but had difficulties keeping her job once her pregnancy became visible. She lived with her Nigerian boyfriend in Kumkapı and hoped that they would get married after the birth of the baby. It is commonly observed that young African women with children experience exclusion from the labour market after pregnancy (Suter 2012: 110-11). At the same time, their vulnerable position renders them more 'eligible' for humanitarian aid and asylum applications (Suter 2013: 112), as discussed below.

Working conditions themselves also function as a mechanism of exclusion. Migrants who lack connections and do not fit the profile for physically demanding jobs are excluded from the labour force. The labour market offers few opportunities for older men and women. Informants have stated that old age and physical fitness are primary reasons for being incapable of finding a job or for their self-exclusion from the labour market. Domestic workers face the difficulty of keeping up with their job as they age. For men who travelled with their families, it gets increasingly difficult to earn enough money to support the entire household as they get older. Among the Afghan families encountered, unmarried (and sometimes married) daughters and school-aged children work, rather than their fathers who are still the heads of the households. Harun made it clear that he and his elder brother earn a living for the family: 'My father does not work, he has gotten old anyway.' The daughter-in-law of another family waiting to cross to Europe explained to me that she and the other children work in textiles, while her father-in-law is more engaged in taking care of the house, cooking, and doing the dishes, noting, 'in my impression, he feels that he does not have a function here.' As in the case of Blessing, middle-aged women are not suited to the pace of production in textile ateliers. Some who are forced to generate income can work as middle persons in textiles, cleaning, and coordinating between different workstations inside the atelier for very low wages. Some also admitted that they needed to send their children to work instead of school. In their study on migrant workers from Azerbaijan, Dedeoğlu and Gökmen (2010: 111) confirm widespread unemployment among middle-aged men and widespread child labour. More attention has been put on migrant child labour in the context of arrival of refugees from Syria.

As discussed above, migrants refrain from explicitly commenting on their experience of neighbourhood violence. Conversely, the immigrants interviewed conveyed more overtly the suffering that stemmed from labour market conditions. Given the exclusionary aspects of the labour market, people employed at the time of the interview expressed their gratitude for receiving an income. However, most informants complained about working long and compulsory extra hours. 'Textile is not as big in Nigeria as it is here. The working hours are not as long in Nigeria. There you work for six hours. Here, it is up to 15 hours. Most guys cannot do it here,' says Peter from Nigeria. Malik, from Afghanistan had worked several years in agriculture in Iran and explains how the hardship of work in Istanbul has impacted his body: 'I have lost a lot of weight since I came to Turkey.' Several people also discussed low wages and the possibility of not being paid or being underpaid. Most find themselves helpless when facing such situations because of their lack of legal status. Victor, a migrant worker from Georgia raises this point:

> There is one very bad thing about people here. You work for six, seven months and they do not pay, they tell you to go away. This is the worst thing [I ask if it has ever happened to him, he hesitates to tell me]. It did not happen to me before as I am a man. If the boss does not pay me I ask using force. I told him, I can force [you] to pay me, he was frightened and he paid me. It happens to women a lot.

Being underpaid or unpaid are particular ways that actors in the market abuse the deportability of migrants. Peter was puzzled by the simultaneous demand for and mistreatment of migrants in the labour market, which left migrants helpless in their employment situations. As explained above, experiences of being tolerated go hand in hand with experiences of being subject to arbitrary abuses and detention. Peter, from Nigeria, has a positive perception of work opportunities in Istanbul. However, he also underlines the efforts required to generate income given the lack of legal and institutional mechanisms protecting migrant workers from the arbitrary practices of employers and security forces. He reported the experience of one of his friends who had to evacuate his workplace without being paid to avoid an alleged police raid:

> The police came and told the boss that they [Africans] should go away. They cannot work because they do not have documents. Sometimes, I ask this question to myself: if a man comes here, if he has nothing to do, why do governments not help this guy? There are jobs that Turkish people cannot do. Sometimes, you need stronger people to carry things. Blacks are stronger in these jobs.

Unlike citizens, migrants of irregular status not only fear losing their jobs, but also deportation. The labour market reinforces this situation because migrants are tolerated as workers participating in the local economy, despite their lack of a formal right to stay in the country. Because their presence is tolerated but not recognized, migrants are left without rights, and very few channels exist to make their voices heard. Indirectly, the possibility of generating an income in the informal labour market contributes to migrants' lack of interest in associating among themselves. Arguably, the image of a docile worker, rather than the rights-claiming activist migrant, fits better in the receiving context of Istanbul, which is characterized by a widespread informal market, low levels of recognition of migrants' rights, and limited institutional/civil societal support for pursuing the rights of irregular migrants. Harsh working conditions make this image of the docile, invisible migrant a reality. Most of them have little energy for engaging in activities outside of their long work hours, let alone political mobilization. Ahmed, 22 years old and from Afghanistan, says:

> I would go [to the association] for language courses, if I did not have a lot of work. In the morning, I start at half past eight, and I barely make it home at half past seven in the evening. Besides, we often do extra hours, three, four days a week, most of the time we cannot go home on time. If there were no extra hours, if I knew that I will be finished by eight every day, I would love to do a course, but with extra hours, it is not possible.

Despite all the hardship inherent in the functioning of the labour market in Istanbul and the lack of access to rights, several migrants who fit the profile of asylum applicants refrain from applying for asylum because they prefer to stay in the labour market. This is often the case for younger migrants, who predominantly chose not to apply due to the availability of work. Those who have already applied frequently decide to live outside of their assigned satellite cities. Common responses from those who prefer to live outside of the asylum system include: 'There is nothing to do in the satellite city. In Istanbul at least you can feed yourself; people, your neighbours can give you food,' and 'I would apply for asylum if they allow me to reside in Istanbul.' The situation implies a trade-off between income generation and the possibility of accessing fundamental rights, as explained in the next section. It also reveals the interconnectedness of recognition and control by the authorities.

Some asylum seekers and recognized refugees waiting to be resettled have faced the dilemma of pursuing their asylum process or generating income through participation in the labour market. Asylum seekers who have lost hope in the asylum process stay in Istanbul at the expense of losing their status. Some asylum seekers are settled in smaller satellite cities near big cities. This arrangement enables them to informally reside and work in one city, but also to commute to their satellite city when necessary to sign in with the police or follow their asylum procedure. Only a small number of asylum seekers can legally reside in Istanbul for medical reasons or special protection needs. Zerrin, a mother of two from Afghanistan, states that the availability of work is the only reason she wants to stay in Istanbul. Zerrin and her two sons applied for asylum months after their arrival in Istanbul. The family needed money to live and pay off debts to smugglers while they waited to reunite with the children's father. To convince the police to allow her to reside in Istanbul, Zerrin had to repeatedly visit the police station to explain her situation. Finally, Zerrin and her sons were allowed to stay in Istanbul, where she could work in the informal labour market:

> They told me that I was too late to apply. I cried a lot. I want to stay in Istanbul because I can work here. I heard that in other cities, there is no work at all. I do not mind about other things. Also, I am used to here. I explained to them that I was here alone with two kids and that my husband was in another country, that I want to live in Istanbul because there is work here, and that there is no work in other cities.

To summarize, this section revealed possibilities for migrants' economic participation in the widespread informal housing and labour market in Istanbul; it also elucidated the limits of labour market participation. Migrants find housing in poorer areas of the city, which are usually close to where they work. These areas are already inhabited by different internal or international migrants and minority groups. It is interesting that even though they live in poor conditions in marginalized areas of the city and are subject to arbitrary controls and abuse by security forces, only a few complained about being subject to neighbourhood crime. Instead, most complaints pertained to the harsh conditions in the labour market and mechanisms of selective incorporation into the labour market that leave groups out of the labour market, including older men – and, to a lesser extent, older women – young women with children, and people with chronic health conditions. Narratives confirm the availability of temporary jobs in the urban economy that require cheap and flexible migrant workers who are 'in good shape', for example in sectors such as textiles, leather, construction, domestic work, and care (see also Toksöz, Erdoğdu, and Kaşka 2012). However, market incorporation comes with a price, as migrant workers are denied fundamental rights. They owe their presence in the labour market to the toleration of the security forces; yet, employers take advantage of their deportability. In the absence of strict internal controls, they refrain from following the expensive and cumbersome work permit process (Toksöz, Erdoğdu and Kaşka 2012: 99-102). Asylum remains a plausible way for migrants to obtain legal status and get access to fundamental rights.

However, the option of asylum is not preferable for most migrants, as they face difficulties in finding comparable jobs in their assigned satellite and therefore would rather stay in Istanbul without any legal status. The blurred distinction between asylum seekers and irregular economic migrants is also apparent in the discussion of fundamental rights in the next section.

4.3   Access to fundamental rights: Between asylum and market

I met Majid, a young man in his early 20s from Afghanistan, in Ahmed's flat. He had arrived in Istanbul after an unlucky journey and needed to be hospitalized upon his arrival. He had walked in extremely cold temperatures for hours while crossing the border from Iran to Van. When he arrived in Istanbul and met his compatriots, he could hardly feel his toes. Luckily, his compatriots knew an Afghan translator working for an NGO who could help Majid fast-track his asylum application. As an asylum applicant, he was admitted to a church-related private hospital, and immediate treatment saved his toes. When I met him a few months later, he was still shocked and not fully recovered, but relieved that his toes and feet would heal. He had not started working yet, as he was not confident that he was ready.

Because of restrictive laws and further limitations in their implementation, irregular migrants' access to healthcare is mainly left to migrants' own means and the extent to which they can afford these services. The principle of universal access to primary education in the law enables children of asylum seekers to go to public schools in their satellite cities. However, children of irregular migrants may be denied formal and informal access. 'Bureaucratic incorporation', namely migrants' access to certain rights and social benefits regardless of legal status as a result of initiatives by street-level bureaucrats, has only been possible for a minority. As in the case of Majid, a closer look at irregular migrants' access to fundamental rights highlights the connection between asylum and irregularity regimes in Turkey.

Opening access to healthcare?

It is essential that applicants, recognized asylum seekers, cover all of their health expenses, themselves.[7]

---

7   From the Implementing Guide of 1994 Regulation, Implementing Guide no: 57, Ministry of Interior, Turkish National Police, 22.06.2006, Retrieved 25.03.2015, from http://www.egm.gov. tr/Documents/uygulama_talimati_2010_genelge.pdf

Access to free, public healthcare for recognized refugees, asylum seekers, and irregular migrants has been an area of negotiation by civil society actors. Recent changes have finally enabled asylum applicants' access to social security, a means-tested system covering access to free or subsidized public healthcare. Firstly, the Law 5510 on Social Security and General Health insurance, enacted in 2008, indicates that asylum seekers and the stateless were included in the general health security schemes (IHAD 2009). Note that the Turkish law differentiates between refugee, asylum seeker, and asylum applicant in the following way: Because Turkey retains a geographical limitation with respect to the application of the 1951 Geneva Convention, as explained in Chapter 2, asylum applicant refers to a person from a non-European country approaching UNHCR and Turkish authorities to seek asylum. The main problem with this procedure has been that the Turkish state does not immediately grant asylum seeker status to recognized refugees (author interview with ASAM, Ankara, December 2012). Despite this limitation, applicants and recognized refugees (even though they do not obtain the status from the state) could apply for healthcare assistance from the Social Assistance and Solidarity Foundations (SAFS) (Şenses 2012: 202-3).[8] However, the availability of aid from SAFS was unpredictable and entailed cumbersome bureaucratic procedures, especially in addressing cases of chronic diseases (author interview with ASAM, Ankara, December 2012). Asylum seekers and migrants of irregular legal status with urgent healthcare needs, and no access to SAFS funds, have become indebted to hospitals.[9] Finally, in response to criticism of the exclusion of asylum applicants from public healthcare insurance, particular articles of Law 5510 were changed in accordance with the LFIP. As envisaged in Article 123 of the LFIP, the expressions of 'asylum seeker and stateless' in Article 3, 27 and 60 of the law 5510 on Social Security and General Health insurance, was replaced with 'person with international protection application, person with asylum seeker status, and stateless person.' Accordingly, the

---

8   Based on Article 1 of The Law on the Encouragement of Social Assistance and Solidarity, Law no: 3294, 29.05.1986 (14.06.1986 official gazette no: 19134).

9   See also press release, by Multeci-der dated 12.11.2013, on the death of an asylum applicant from Afghanistan, suffering from a chronic kidney problem. Reportedly, the patient did not receive sufficient financial aid from the state departments or from UNHCR and eventually refused dialysis treatment, as her family was heavily indebted to the hospital. The case reveals the limitations of access to healthcare even for applicants under the international protection regime. See *Tajik'in ölümü* ('the Death of Tajik'), *MÜLTECİ-DER Press Release* 12.10.2013. Retrieved 25.03.2015, from http://www.bianet.org/system/uploads/1/files/attachments/000/000/981/original/Multeci-Der_Tajikin_olumu.pdf?1381755437

new legislation included asylum applicants, along with asylum seekers, stateless, and recognized refugees in the scope of general health insurance.

The state's recognition of its responsibility towards persons seeking international protection has been a positive development. Whether these changes will ensure applicants' access to health services without the intervention of NGOs is yet to be seen. Based on their experiences with hospitals, most NGOs have been cautious in celebrating this legal change. Even in the case of Syrians, who are under the temporary protection regime, access to healthcare in Istanbul has been reported as problematic.[10] Most NGOs working on healthcare complained about the arbitrariness of street-level bureaucrats in the functioning of the healthcare system (regardless of the patients' legal status): 'These circulars do not work automatically, somebody has to push. [...] It is not only about the hospital, it depends on who is on the shift' (author interview with MSF, Istanbul, November 2013). In this sense, it is difficult for asylum seekers to directly access hospitals without an NGO acting as an intermediary, and they are forced to negotiate almost each case from scratch.

Improvements regarding asylum applicants' access to healthcare have arguably reinforced the distinction between asylum applicants (conceived as needy refugee) vs. migrants with no legal status (conceived as illegal). According to current legislation, irregular migrants may have access to healthcare or other social assistance services only when they are detained or identified as victim of trafficking[11] (Şenses 2012: 201-3), or if they become asylum applicants. However, the asylum applicant status does not bring automatic access to free public healthcare, for the reasons stated above, because one needs to be registered and acquire a foreigner's ID number to get free access to hospitals (Balta 2010: 38). In urgent cases, such as Majid's, encouraging patients who fit the asylum seeker profile to make an asylum application is one way that civil society actors highlight to ensure migrants' access to healthcare. As in the case of Alima, church-related organizations help pregnant women get access firstly to asylum procedures and subsequently to hospitals. The LFIP, while recognizing

10   According to the circular on healthcare and other services provide to Syrian Guests, published on 09.09.2013, by the Turkish Prime Ministry Disaster and Emergency Management Authority (AFAD), all Syrians registered with the authorities have access to free public health care.

11   See the Circular on the Provision of aid by SAFS to non-citizens and foreigners in vulnerable conditions, General Directorate on Social Aid and Solidarity, Circular no: 8237, 20.05.2009. Retrieved 25.03.2015, from http://sosyalyardimlar.gov.tr/mevzuat/genelgeler/yardimlar-dairesi/20052009-tarihli-ve-8237-sayili-genelge

asylum applicants' right to welfare services, is cautious about the use of the asylum system for getting access to healthcare. Article 89, Clause 3 envisages financial benefits for those who applied for asylum to get access to free healthcare:

> For those applicants or international protection beneficiaries who at a later date would be found to already have had medical insurance coverage or the financial means or, to have applied [for asylum] for the sole purpose of receiving medical treatment shall be reported to the Social Security Authority within ten days at the latest for termination of their universal health insurance and the expenditures related to the treatment and medication shall be reimbursed from them.

This measure to prevent bogus asylum seekers from accessing free public healthcare through asylum reveals the political will to distinguish economic migrants from 'genuine refugees'. Again, the implementation of this precaution is yet to be seen, as it would be difficult to prove that a person does not have genuine asylum claims.

Given the laws and practices restricting irregular migrants' access to healthcare, NGOs focusing on the physical and mental well-being of asylum seekers and migrants generally provide free basic consultations. Common diseases among migrants mostly stem from living conditions, lack of hygiene, infections, and psychological problems resulting from the long journeys that some had to take (author interview with TOHAV, November 2013, Istanbul). While consultations are accessible to all migrants, regardless of legal status, the possession of certain papers, such as an asylum application or passport with valid entry, may be necessary to negotiate access to hospitals for secondary level treatment or analysis. Association for Solidarity and Mutual Aid with Migrants (ASEM), an NGO running a small clinic initially funded by Medecins du Monde in the Kumkapı area explained that patients in need of secondary treatment can be taken to hospitals, but even church-related hospitals, known to be more open to migrants, require certain documents (at least a passport). Another option for negotiating migrants' access to hospitals is through emergency rooms, which lowers the costs (author interview with ASEM, Istanbul, November 2013). Again, the latter strategy is more likely to work for legal entrants. These efforts are very limited, given the human and financial capacity of these civil societal institutions.

In the absence of access to free public care, migrants of irregular status and NGOs providing humanitarian aid have had to cover the cost of healthcare. The introduction of tourist fees in 2011 has worsened access to healthcare, as this increased the cost for migrants without legal status. The circulars introduced by the Ministry of Health in 2011, updated in 2013, on Health Tourism and the Provision of Services in the context of Tourists' Health requires higher fees from foreigners without residence permits, including tourists.[12] The circular exempts asylum seekers, applicants, victims of human trafficking, and administrative detainees, but applies to migrants of irregular legal status. A Syrian migrant of Turcoman origin was able to acquire a residence permit, but was puzzled by the introduction of tourist fees when he visited the hospital before and after the legal change: 'It was 15 TRY a week ago and 100 TRY a week after. This is very difficult. If the person does not have social security, the person would die out of hunger.' Tourist fees, applied to immigrants who are not tourists, not only reveal that migrants have become victims of the general marketization of the healthcare system in Turkey (see Agartan 2012) but also how migrants without legal status are forced to avoid medical help until they are in dire need and how they are left at the mercy of the market. As a result, migrants are left unprotected against health risks, and healthcare is only available to those who can afford it.

Given the limitations of institutional support in terms of getting access to free public healthcare, most migrants rely on their own resources or their community/friends' networks. In most cases, private health clinics are 'chosen' over public ones. When Natalia, a domestic worker in her mid-forties from Moldova, living in and out of Turkey for over 10 years, needed to see the doctor, her boss took her to a private consultation and covered the expenses. Especially for those without papers, going to private clinics and hospitals reduces the chances of rejection and the risk of revealing oneself as illegal. Ahmed from Afghanistan had no papers to prove his identity when he had appendicitis. After being rejected by a public hospital, he received treatment from a private one with the help of one of the Afghan associations in Zeytinburnu:

[Referring to the White Passport received from the Afghanistan Consulate], I did not have a passport then. One day, I left work, went to an

---

12   See, the Circular no 2011/41 on Health Tourism and on the Provision of Services in the context of Tourists' Health, *saglikekonomisi.com*, 15.06.2011. Retrieved 25.03.2015 from http://www.saglik-ekonomisi.com/sed/index.php/haberler/446-saglik-turizmi-ve-turistin-sagligi-genelgesi Updated Circular dating 23.07.2013, approved by decision no: 25541. Retrieved 25.03.2015, from http://www.saglik.gov.tr/DH/dosya/1-88061/h/saglik-turizmi-ve-turist-sagligi-kapsaminda-sunulacak-s-.pdf

internet [café]. I was on the internet for one hour or so, then I had pain in my belly. I went home, thought the pain was normal and did not go to hospital until the evening. Then, I told my friend, who took me to hospital. It was appendicitis, I was given a serum and the pain was gone. Then, we went to the big hospital. I had nothing with me, no passport, nothing. The hospital rejected me. Then, we came back home. The head of the association in Zeytinburnu called the XX medical centre [a private hospital in the neighbourhood]. I was only then admitted. [...] I had two more serums. I would go through a surgery if the pain had come back. I was ok after the two serums [...] I had to pay all the cost myself. I spent almost 1000 TRY. I had little on me, the rest I borrowed from friends. Everybody had brought money for me because I went to hospital.

Ahmed never went back to the hospital after this incident. He went to Ankara to apply for a white passport, and he knew that 'the passport is only valid if the police stops you, then you show the passport, it is no good for hospital or anything else. [...]. If you have a residence, then it might be different.'

As Ahmed's case shows, the reliance on the market for healthcare goes together with a reliance on ethnic networks for informal consultations, and, at times, on alternative forms of healing. Unsurprisingly, migrants first seek help within their communities. Harun's mother, who later joined him, had a cataract operation in one of the private hospitals known to have a formal agreement with one of the Afghan associations. Afsana, a mother of three also from Afghanistan, could not endure working conditions after a couple of months of working in textiles in Istanbul. She was complaining about back pain and paid for her scans at a private clinic. As the family could not afford a consultation with a specialist, they waited for a visit to an Afghan doctor, who had acquired Turkish citizenship. The doctor could not say much by only looking at the scan, but told her that she had a lot of pressure on her back and that a specialist could prescribe her medicine and an exercise programme to follow. In the absence of such informal consultations, alternative healing methods may be the only option. One Sunday, for instance, after a ceremony in an African church near Taksim, Alima received a bottle of olive oil from the pastor to apply to her legs. She had severe, possibly bone-related, pain in her legs, but could not afford to go to the doctor despite her asylum applicant status.

Over the last decade, access to healthcare has moved from the exclusion of migrants, asylum seekers, and applicants alike, to the recognition of the healthcare rights of those included in the international protection regime. In the absence of plausible legal grounds, the inclusion of so-called economic migrants in the existing public healthcare scheme is only possible through asylum applications. Conversely, by envisaging sanctions for the use of international protection to get access to healthcare, the new law reinforces the legal distinction between asylum seekers and irregular migrants. Most NGO activities are channelled to providing basic services and enabling migrants to receive urgent access to healthcare through asylum, if possible. Given the limits of such efforts and the arbitrary practices in hospitals, migrants of irregular status rely on their own financial resources and communal networks, and hence, are largely in the hands of a highly privatized healthcare system.

Education

Sima was eight years old when I met her. Her family was originally from Afghanistan and had been in Istanbul for a year, after staying in Iran for several years. Sima and her younger brother, Nader, cannot go to school because the family crossed the border without the necessary papers and settled in Istanbul. Sima's parents knew that the children could go to school if they applied for asylum. The family was reluctant to do so, however. They knew from their relatives that the process was especially long and inconclusive for Afghans, and they would have to go and live in a remote city. Sima's mother, Afsana (37), explained her main concern, 'it is only for the kids, so that they can go to school.' The family had gone through hard times when I first met them. The eldest daughter, also of school age, and the mother, despite her health problems, were working in textiles, and the father was unemployed, while Sima looked after her brother and undertook household chores such as cleaning and cooking. The situation deteriorated when they were expelled from their flat in the aftermath of flooding and fire incidents. When I met them six months later, they were more settled in another flat in the neighbourhood. The father had found temporary work in construction during the summer. Although no solution was found for the children's schooling, asylum was no longer a viable option, since it would mean sacrificing income-generating possibilities that had become necessary for the survival of the household.

According to the Turkish Constitution Article 42 and legislation on access to primary education, primary education is compulsory in Turkey for both citizens and foreigners, with or without legal status.[13] In addition, access to free public education is a right, as stated in the Convention on the Rights of

---

13   See the Law on Primary Education, Law no: 222 05.01.1961.

the Child that came into force in Turkey in 1995. Despite these legal measures, schooling of children of irregular migrants has been characterized by exclusion, self-exclusion, and informal inclusion. Access to primary education is less problematic for the children of asylum seekers in satellite cities (author interview with HCA, Istanbul, November 2013). However, because of bureaucratic exclusion, or the 'whims of bureaucrats' in Kitty Calavita's terms (2005: 108), the children of irregular migrants and asylum seekers not residing in the assigned satellite city may not get formal access to public schools. The best scenario is to have informal access to schools through negotiations with the bureaucracy. Formal access to schools is tied to residence permits, and for many families, access to residence permits is only possible through the asylum application procedure. The functioning of the asylum procedure does not offer prospects for resettlement to a third country or a permanent status in Turkey. Given this obstacle, many families, such as Sima's, are torn between informal labour market opportunities in Istanbul and enjoying fundamental rights, including children's schooling in satellite cities.

The lack of a residence permit is the primary obstacle to children's formal access to public schools (Danış, Taraghi, and Pérouse 2009: 627-8). In line with previous research, the interviewed Afghan families without residence permits underscored the continuing problem of schooling for Afghan children in the Zeytinburnu area (see also Toksöz, Erdoğdu, and Kaşka 2012: 123). For Harun's family, one of the main motivations to apply for a residence permit through their relatives in Antakya was the possibility of schooling for her younger sister: 'The lack of residence is very difficult for children. My sister is crying all the time. She is supposed to start the third grade this fall. It has been two years, she cannot be enrolled in the school, [because we have] no residence.' Given the reluctance of school principals to admit students without papers, even informal enrolment becomes a privilege. At the discretion of the principal, some children can be admitted as guest students and follow courses without receiving formal degrees (author interview with ASEM, Istanbul, November 2013). Minority schools, such as Armenian schools, are known to accept undocumented children 'as guest students' (EC 2014: 61). Meanwhile, Sima's parents tried all of the primary schools in the area and were rejected several times. In the absence of opportunities for formal or informal enrolment in public schools, children attend temporary courses provided by church-related NGOs or migrant associations. 'We tried two times, and we were told that we do not have a residence permit. Now, she is going to school, but only two to three times a week, after school hours. The teacher helps them to learn how to read and write,' Harun explained.

Self-exclusion may be the case for families who either give up on the idea of sending their children to schools or do not try in the first place. Some families who intend to leave the country do not find it necessary to send their children to school in Turkey. Arriving in Turkey with the intention to cross, their stopover in Istanbul is primarily aimed at collecting the money needed to pay smugglers. For families hoping to cross to Europe or to reach a third country through resettlement, children's contributions to the family's income may be more important than their education in Turkey. Echoing the observations made by Danış, Taraghi, and Pérouse (2009: 573) in the case of Afghan families in Zeytinburnu, Zerrin, for instance, hoped to resettle and join her husband in the Netherlands. Therefore, she was not concerned that her sons did not attend school and worked in textiles instead: 'I only want a course for them, to study English, but there are no such things here. Private courses are expensive. There are state courses, but only those with a residence permit can go there.' For Sima's family, on the other hand, who lacked the legal and financial resources to go to another country, the requirement of asylum procedures to reside in a remote city was the main reason they were reluctant to apply for asylum. Sima's mother, who had completed secondary education, used to do clerical work in Afghanistan and was puzzled by the question of schooling of her children: 'As long as we can make a living, it would not matter to me to go wherever. It is only for the children, so that they can go to school. For us, it is too late anyway, but kids, they must go to school.'

Earlier research has revealed the difficulty of getting access to basic services such as healthcare and education without an official status (Danış, Taraghi, and Pérouse 2009: 627-8; Şenses 2012: 204). Irregular migrants' healthcare needs are left in the hands of the market (as also observed by Danış, Taraghi, and Pérouse 2009: 627). What has received less attention, however, are the practical implications of the legal distinction between irregular migrants and asylum seekers. The gradual recognition of certain rights in favour of asylum seekers, despite implementation problems, has had consequences for the rights of irregular migrants as well as for civil society practices. For migrants who fit the asylum applicant profile, such as those from Iran, Iraq, Afghanistan, Pakistan, and African countries, applying for asylum has been a shortcut to obtaining legal status. Asylum applicant status gives them and their advocates a degree of legitimacy in negotiating rights for forced migrants.

To conclude, the mechanisms of bureaucratic incorporation, in the sense that irregular migrants can be 'regularized from below', as a result of citizenship practices at the grassroots (Nyers and Rygiel 2012: 15), by getting legitimate access to certain rights, i.e. by sending their children to school regardless of their lack of legal status, has been very limited in the case of Turkey. Civil society activities and claims based on the narrative of forced migration, i.e. refugees, have arguably pushed the state to respond to these critiques by recognizing certain fundamental rights of asylum seekers and refugees. Conversely, such legitimate grounds have not been carved out for the rights of irregular economic migrants. Irregular migrants have found fewer channels to express their marginalization. The absence of institutional support and protective measures reinforces their vulnerable position in society, comprising a cheap and available labour force in the informal market, as discussed in Section 4.2. The realm of advocacy has reinforced this legal separation and arguably precluded the emergence of a more comprehensive political movement for the rights of migrants in general, as discussed in the case of Morocco.

4.4   Reversing illegality: Mobilization or moving sideways?

The objectified person "is seen but he [sic] does not see, he's the object of information, never a subject in communication. (Foucault 1977: 200 quoted in Shore and Wright 2003: 4)

On 2 September 2014, a young Congolese man was murdered in his house, in the Tarlabaşı neighbourhood,[14] very close to Taksim Square.[15] On 8 September 2014, a group of African migrants living in Istanbul organized a press conference on Istiklal Avenue, a prominent area for public protests, opposing 'racist murders', with the support of pro-migrants' rights associations. The banner used in the press conference said: 'Africans and migrants are not alone. Stop racist murders.' The meeting also underscored that migrants could not go to the police when they were targeted by racist crimes because of their fear of deportation. Unfortunately, both the crime and the protest received little media attention. Moreover, the media depicted the murder as a case of homicide, rather than as a racist or xenophobic attack.

14   Despite the ongoing gentrification, the area has been inhabited by internal migrants from eastern parts of Turkey, largely populated by Kurds, and international migrants. It is one of the neighbourhoods where migrants from the African continent are most visible.

15   *Beyoğlu'nda korkunç cinayet* ('Terrifying murder in Beyoğlu'), Gazete Vatan, 02.09.2014. Retrieved 10.09.2014 from http://www.gazetevatan.com/beyoglu-nda-korkunc-cinayet-675773 -yasam/

This protest was one of the few, exceptional instances in which migrant communities residing in the city have made themselves visible in the public sphere and contested xenophobic violence. This section connects with the earlier discussion on the arbitrary practices of subordinate forms of inclusion and exclusion. Underscoring the rarity of such street protests by migrants and pro-migrant actors, this section first shows the specific institutionalization of civil society around asylum issues in Turkey, which undermines issues pertaining to the rights of irregular migrants. Secondly, it explores the fact that migrants of irregular legal status have had fewer opportunities to raise a political voice in the absence of institutional support. Due to this lack of institutional support, most migrants interviewed in Istanbul linked their prospects for legal status to individual or ethnicity-based legalizing efforts. The latter may be possible for those who can prove that they are of Turkic ethnicity (Danış and Parla 2009).

Civil society working on immigration issues

As discussed in Chapter 2, the emergence of civil society organizations centred on migration and asylum issues in Turkey is rather recent. Its institutionalization dates back to the heydays of the EU accession process in the pre- and post-2005 period. Human rights organizations such as the Helsinki Citizens Assembly (HCA), Association for Solidarity with Refugees (MÜLTECİ-DER), the Association of Human Rights and Solidarity for Oppressed People (Mazlumder), Amnesty International, and the Human Rights Research Foundation (IHAD) engage in advocacy and/or provide legal aid, mainly to asylum seekers. Rights-based institutions working in the field of asylum and migration formed the Commission for Refugees in 2010. There are also organizations that provide humanitarian aid and services to migrants and refugees, including the Association for Solidarity with Asylum Seekers and Migrants (ASAM), the Human Resource Development Foundation (HRDF), Doctors without Borders (MSF), Foundation for Society and Legal Studies (TOHAV), ASEM, church-based organizations such as Caritas, and the Istanbul Inter-Parish Migrants Program. Among these, some have been implementing partners of UNHCR and organizations such as ASAM, HCA, TOHAV, MÜLTECİ-DER, and ASEM, and have benefited from EU and other international funding. With the arrival of Syrians, existing civil society actors such as Support to Life developed an interest in the subject and new ones emerged. Furthermore, Islam-oriented charity organizations, such as the Humanitarian Relief Foundation, expanded their activities to include the field of asylum.

Grassroots networks such as the Migrant Solidarity Network (GDA), an activist network emerging in 2009 around the idea of the unconditional right to mobility, show solidarity with migrants regardless of legal status. There are similar networks which are not directly organized around issues of migration, but whose sectoral focus concerns irregular migrants in the labour market, such as *Ev İşçileri Dayanışma Sendikası* ('Union for Solidarity with Domestic Workers'), *Geri Dönüşüm İşçileri Derneği* ('Association for Recycling Workers') (Toksöz, Erdoğdu, and Kaşka 2012: 117), and *İşçi Sağlığı ve İş Güvenliği Meclisi* ('Assembly for Workers' Rights and Work Security'). Apart from civil society and social network organizations that work on immigration issues, formal trade unions have been remarkably inactive on the question of irregular migration (Şenses 2012: 215-6; Toksöz, Erdoğdu, and Kaşka 2012: 128). Although they recognized the fact that Turkey has increasingly been receiving labour migrants (see TES-İŞ 2005), they have not embraced a pro-migrants' rights stance. One explanation for this is because major trade unions in Turkey are organized in the formal economy, while most migrants work in the informal economy.

Research has also observed that the focus of civil society activities in Turkey has been on asylum seekers rather than irregular migrants (Ozcurumez and Şenses 2012: 90, 104; Şenses 2012: 210; Parla 2011: 82; Balta 2010: 105). Civil society has prioritized the protection needs of asylum seekers, even though there were few asylum seekers until 2011, compared to the estimated number of irregular migrants in Turkey in the early 2000s. Despite this general trend, a closer look at the humanitarian practices of civil society actors reveals that they do not totally exclude irregular migrants, as discussed in Section 4.3. Meanwhile, the advocacy activities of civil society underscore human rights violations in relation to asylum. The NGO reports have mainly revealed malfunctions in the asylum system in terms of access to asylum procedures and asylum applicants' access to fundamental rights (IHAD 2009; AI 2009; HRW 2008). The analysis of such reports and civil society press releases reveals almost no references to irregular migrants, their labour market conditions, or their access to fundamental rights. Documents generated by civil society usually refer to irregular migrants in detention as potential asylum seekers who cannot get access to international protection procedures.

One exception in the language of advocacy has been in the work of civil society's provision of healthcare, with the initiation of claims for all migrants' access to health, regardless of their legal status. The press release organized by ASEM, an NGO based in Kumkapı, Istanbul, providing direct healthcare and consultation to migrants, in collaboration with other civil society organizations working on healthcare on International Migrants' Day, 18 December 2014, contested the marketization of healthcare and hierarchies stemming from legal status that inhibited irregular migrants' access to legal status.[16]

When asked about their focus on asylum issues, most NGOs acknowledged that they have to prioritize these, not because they are insensitive to irregular migrants' issues, but because they lack the capacity, resources, and expertise to extend aid to all migrants. 'Our expertise now is on asylum procedures, but we also follow policies in the field of migration in general,' explained an informant from HCA. The argument on the lack of expertise is generally motivated by a lack of financial and administrative capacity to cover issues pertaining to irregular migration and the inability to reach this diverse population in terms of legal status and protection (also suggested by Balta 2010: 107). The informant from Amnesty International explained:

> Definitely, it is a matter of resources. Immigrants in Turkey is a huge area, the numbers may reach millions. Working on migrants cannot be limited to undocumented migrants. One needs to include domestic workers, even students, those who come to work, overstayers. There is no organization big enough to undertake this. Even in the area of refugees, many organizations are limited by their lack of resources. Among all refugees in Turkey, how many of them are aware of NGO activities? Many have not even heard about them. With projects they undertake and resources they receive, NGO services are very limited. It is not sufficient to reach all 30,000 refugees in Turkey. The area of immigration is much bigger, NGOs would need a huge amount of resources and much bigger capacity. (Author interview with Amnesty International, Ankara, November 2012)

The issue of access mentioned in the quote is also revealed as a general obstacle even though some NGOs provide services to all migrants regardless of legal status. Asylum seekers arguably have more knowledge of formal organizations that provide support than unregistered migrants. As articulated by an NGO, 'As a charity organization, we cannot reach people, several people pass without touching any institution' (author interview with CARITAS, Istanbul, January 2014). Irregular migrants are segmented into different sections in society and are hard to reach. As a result, most of the

---

16   See, Migrants cannot access health care services, 18.12.2014, *guvenlicalisma.org*. Retrieved 10.09.2014 from http://www.guvenlicalisma.org/index.php?option=com_content&view=articl e&id=12560:gocmenler-saglik-hizmetlerine-erisemiyor&catid=152:haberler&Itemid=141

clients of these humanitarian services are asylum applicants. Given their limited resources and capabilities, NGOs prefer to distinguish between urgent and less urgent cases. This attitude implies prioritizing the needs of potential asylum seekers, hence it reinforces the distinction between political refugees and economic migrants. As articulated by Mazlumder, in response to how they evaluate NGO activities on irregular migrants:

> Let me put it this way. The cases coming to us are more urgent cases, if it is appropriate to put it this way. It is like an emergency room in hospitals. Refugees who do not have legal problems do not reach us or do not have the need to reach us. Those under pressure of deportation come to us. People facing the danger of deportation to countries where they will face persecution is the exact area we study. We actively cover cases like this. For regular, irregular migrants, as far as I know, humanitarian aid organizations may help. (Author interview with Mazlumder, Istanbul, November 2013)

With a rapidly increasing caseload, it is likely that asylum-related issues will continue to dominate civil society's scope. This dominance, however, is more linked to the character of the emerging field of governance, where UNHCR and EU's concerns prevail. The civil society actors interviewed acknowledged that most of the regulations concern asylum seekers and leave migrants from other categories to their own fate. Arguably, UNHCR's central role within this governance regime and the bureaucratic routine institutionalized by UNHCR influence the activities of civil society (Scheel and Ratfisch 2014: 928; Balta 2010: 106). My findings resonate with this observation. On the one hand, irregular migrants are, to a certain degree, criminalized, especially in their relation to human and drug smuggling networks. On the other hand, irregular migrants' access to rights has been sidelined by civil society actors in the field. Consequently, the issue of irregular migrants' access to rights is further depoliticized by civil society activities.

The generally low levels of the politicization of irregular migration, discussed in Chapter 2, reinforce UNHCR's dominance in the fields of migration and asylum in Turkey. Given this particular shaping of the field, humanitarian NGOs' focus on asylum has provided a more legitimate basis for expressing their mandate, although most humanitarian NGOs do not explicitly distinguish between refugees and irregular migrants especially in practice. In contrast to previous research that implies a total ignorance towards issues pertaining to irregular migration (Şenses 2012: 205), I suggest that the focus on asylum is more than a strategic use of limited resources.

As explained above, doing advocacy in the name of asylum seekers has become a legitimate way for NGOs to express their concern with human rights violations in Turkey.

Despite this general inaction on irregular migration, NGOs have vividly contested arbitrary detention and deportation practices that concern irregular migrants and asylum seekers alike. Such contestations have given rise to a vivid battleground for NGOs to ensure the rights of migrants who are trapped in irregular situations. Civil society has also functioned to stop unlawful deportations through interim measures taken by the ECtHR (Grange and Flynn 2014: 24; Yilmaz 2012; Ulusoy and Kılınç 2014). Ulusoy and Kılınç (2014: 255) emphasize that lawyers from Turkey have made less use of the ECtHR's interim measures than their counterparts in other European countries. Meanwhile, civil society has used the ECtHR's decisions to oppose arbitrary detention and deportation decisions taken by security forces (Yilmaz 2012: 51-2). The Court took exemplary decisions against Turkey, acknowledging that asylum seekers do not have access to procedural guarantees in the country (Ulusoy and Kılınç 2014: 255). Furthermore, detention and deportation practices were ruled to be in violation of Articles 3, 5, and 13 of the ECHR (Grange and Flynn 2014: 19).[17] Turkish NGOs' critiques and their use of the ECtHR as a transnational accountability mechanism have largely contributed to the preparation of the LFIP (Şenses 2012: 218-9). The law has ensured that the practices of deportation and detention are in line with the requirements of the ECHR (Yilmaz, 2012: 54-5).

In response to widespread international and domestic critiques, the asylum regime saw improvements before the law came into force. The cumulative creation of law led to increased access to the asylum process in detention. Both HCA and Mazlumder have noted a relative improvement since 2010 in terms of access to asylum after detention and that officials have become more prone to accepting the asylum applications of detainees, especially in Istanbul, rather than releasing them with deportation papers. The decreasing number of deportations from Istanbul and increasing number of asylum applications from 2010 to 2011 reveal that the Turkish police have become more inclined to channel detainees towards asylum procedures (personal communication with HCA).

---

17   See for instance Abdolkhani and Karimnia v. Turkey, Z.N.S vs. Turkey, Charahili v. Turkey 'that concluded the absence of clear provisions for ordering and extending detention, the lack of notification of the reasons for detention and the absence of judicial remedy to the decision on detention and torture.' Ranjbar and others vs. Turkey, sentencing Turkey for unlawful detention (Yilmaz 2012: 162,) Db vs. Turkey, sentencing Turkey for unlawful deportation July 2010 (Yilmaz, 2012: 169).

Legal sidesteps in the absence of mobilization

What are the implications of these advocacy activities for the rights of irregular migrants? As discussed in Chapter 2, in terms of advocacy, deportation and detention practices, asylum applicants' access to fundamental rights, and access to asylum in general have been major areas of contention. This legal activism in the area of deportation also reflects the limitations in the sphere of intervention by civil society. Migrant illegality is, to a certain extent, reversed, but this occurs by turning migrants into asylum clients rather than through activism for irregular migrants' access to legal status. Such attitudes reinforce the idea that asylum is the only way to obtain recognition and hence legitimacy, not only in the eyes of state authorities, but also at the level of advocacy. In this sense, rather than engaging in protesting the deportations of migrants for humanitarian reasons, the contestations remain within the limits of the law. This self-limitation unintentionally led to the depoliticization of the detention and deportation of migrants who do not fit the asylum seekers' profile in Turkey.

The primary form of contestation by civil society actors in Turkey has been legal activism rather than street protests. Street protests, as exemplified at the beginning of the section, are organized in a sporadic fashion. For instance, the suspicious death of Festus Okey, a Nigerian asylum seeker, in police custody in 2007, and the process of his trial, propagated a series of street protests as well as media and public attention.[18] Protests organized outside detention centres in Kumkapı, Istanbul, and Edirne contested the unlawful detention and deportation practices of the state in particular, and immigration and asylum policies in Turkey, in general.[19] Such events

18   See for instance, HCA Condemns Refugee Death in Police Custody. *bianet.org*, 31.08.2007. Retrieved 15.03.2014, from http://www.bianet.org/english/human-rights/101489-hca-condemns -refugee-death-in-police-custody

İHD: Festus Okey'in Öldürülmesini Protesto Etkinliğine Davet ('Human Rights Association's (IHD) call to protest the murder of Festus Okey'), *savaskarsitlari.org*, 03.09.2007. Retrieved 15.03.2014, from http://www.savaskarsitlari.org/arsiv.asp?ArsivTipID=9&ArsivAnaID=40719

Police Cover Up in Okey's Death, *bianet.org*, 13.09.2007. Retrieved 15.03.2014, from http://www. bianet.org/english/human-rights/101739-police-cover-up-in-okeys-death

19   See for instance, Call for Action in support of immigrants, 23.09.2009. Retrieved 15.03.2014, from https://resistanbul.wordpress.com/2009/09/23/call-for-action-in-support-of-immigrants. This was a street protest in solidarity with detainees' uprisings in Kumkapı Detention Centre, as part of anti-capitalist protests during the International Monetary Fund and World Bank Summit in 2009, in Istanbul.

Participants of the international *Transborder Conference* attended a protest at the gates of Edirne Detention Centre that took place in March 2012 in Istanbul. Protest at the Detention Centre in Edirne, Turkey: The border is the problem!, *wreu.info*, 21.03.2012. Retrieved 15.03.2014, from

triggered coalition building among CSOs and attracted civil society allies. However, their potential to include migrants as well as asylum seekers as active rights-seeking agents has remained limited. Sit-ins and hunger strikes by Afghan refugees that took place in Ankara for over a month in April and May 2014 were exceptional protests that saw refugees themselves at the forefront of action. However, the target of the protest was UNHCR, which had suspended asylum applications from Afghan nationals, rather than the Turkish state.[20]

One reason for lower levels of street activism by migrants is that there are few channels available for the political mobilization of irregular migrants and refugees. As reported by an IOM-funded report, existing migrant organizations are not powerful enough to raise their voice (Toksöz, Erdoğdu, and Kaşka 2012: 113-4). They are either ethnicity-based recognized associations, founded by migrants who arrived within the context of the Settlement Law and have acquired citizenship, or small-scale community/ethnicity-based solidarity groups. The capacity of older, more established ethnic associations to lobby for newly arrived irregular migrants has been limited and selective (Danış and Parla 2009 155-6; Parla 2011). Research has shown that ethnicity-based informal solidarity networks among African migrants have been short-lived because members tended to be highly mobile, and 'the transit matter inhibits solidarity' (Suter 2012: 208). Resonating with these observations, my findings show that there are no migrant associations crosscutting ethnic differences in Turkey. This is mostly due to the diverse profiles of migrant communities in terms of ethnic, linguistic, and even cultural backgrounds. Moreover, migrants in irregular situations show low degrees of mobilization, even within the same ethnic group. Internal differences within one ethnic group need to be taken into account rather than taking ethnicity as the 'key mobilizing category' (Però and Solomos 2010: 9). In this sense, the Union of the Young Refugees in Turkey (UJRT, abbreviated from the French name for the organization), formed in 2010, has been an exceptional example of inter-ethnic solidarity among refugees. The association has worked as a solidarity network to improve the living conditions of refugee minors who had to leave their state and runs shelters for those who were left to fend for themselves when they turned 18. They

http://infomobile.w2eu.net/2012/03/21/protest-at-the-detention-center-in-edirne-turkey-the
-border-is-the-problem/

20 With mouths sewn shut, Afghan refugees keep protesting Ankara, UNHCR, *hurriyet-dailynews.com*, 26.05.2014. Retrieved 15.03.2014, from http://www.hurriyetdailynews.com/with-mouths-sewn-shut-afghan-refugees-keep-protesting-ankara-unhcr-.aspx?pageID=238&nid=67005&NewsCatID=339

forged close alliances with international and civil society organizations such as IOM, UNHCR, the Migrant Solidarity Network, and Caritas. Although the resettlement of their members into third countries has been a priority for UJRT, they also mention integration into Turkish society as an objective. These initiatives can potentially evolve into other forms of activism that might include other groups of migrants and refugees. However, for the reasons stated, mobilization has remained limited.

In the absence of bottom-up demands for regularization, in the summer of 2012, the one-time amnesty initiated by the Ministry of Interior presented a unique legal opportunity for illegal migrants to regularize their status. Note that this was a top-down measure aimed at registering and reducing the number of clandestine workers and those overstaying their visa, rather than a response to civil society's or employers' formal demands.[21] The amnesty enabled migrants whose visa or residence permit had expired to get access to a residence permit for six months. Accordingly, overstayers who held valid passports, agreed to pay fines for the time they overstayed, and those able to show a valid rental contract or sponsor letter were granted a one-time exceptional residence permit. Those who did not want to extend their residency were invited to leave the country by the end of 2012. Several migrants, including Chris, a 36-year-old man from Nigeria, who agreed to pay fines to regularize their status, were later left feeling deceived: Chris's plan was first to regularize his status and then to continue his graduate education at a private university in Istanbul. Chris used his six-months residence permit to travel back and forth to Nigeria in order to secure money from his family for his education. Ultimately, his enrolment at the university was not possible due to bureaucratic problems, and he again found himself with the overstayer legal status, just like before the amnesty: 'The *ikamet* ['residence permit'] was just good for leaving and coming back. It never helped with anything else.' Similarly, Peter, another young man from Nigeria, also frustrated with the amnesty law, reiterates its economic aspects: 'I do not call it an amnesty, I call it a robbery. When you give a residence permit that you cannot renew, it is robbery. I cannot say that now I have documents. They are invalid.' For many, there was a similar discrepancy between the cost of legalization papers and their benefits.

---

21   See Ministerial Approval no: 108807, Residence Regulation for Foreigners having violated visa/ residence in Turkey. Retrieved 15.03.2014, from http://eng.yabancicalismaizni.com/services/ residence-permit-in-turkey/297-new-regulations-have-been-made-to-make-visa-work-and-residence-permits-of-illegal-foreigners-easier.html)

Turkic ethnicity is another basis for acquiring legal status in Turkey (Danış and Parla 2009; Toksöz, Erdoğdu and Kaşka 2012: 113-4). However, low levels of trust characterize irregular migrants' relations to ethnicity-based cultural organizations established by earlier immigrant groups who have acquired citizenship (Parla and Kaşlı 2011; Kaşlı 2016). Among the interviewees from Afghanistan, Selma, a Farsi-speaking migrant, mentioned her perception that Afghan associations are divided along ethnic lines and are less interested in helping Afghans from other ethnicities: 'We never went to the association. There are associations by Kazakhs, Uzbeks here. Those who are here, only help Uzbeks, for instance, they do not provide help for other migrants, only Uzbek. This is why we do not go to the association.' Malik, an Afghan migrant of Uzbek ethnicity explained that he was reluctant to go to the association in their neighbourhood because he never received proper answers to his questions on residence permits:

> I went there once. The person there did not talk to me properly. I asked, 'isn't this association for foreigners, I am here to ask my questions, why you do not answer correctly?' The person told me there are too many people out there asking for help and they cannot answer all the questions thoroughly.

Malik says he left the association frustrated and did not trust the person who later asked him for money to get him a residence permit:

> He asked for 100 dollars and told me he will get me a residence permit in six months. I told him that I would give him 200, but only after he got get my permit. He called a few days later. I gave him my passport [white passport taken from the Afghan embassy, not a travel document], he made copies. He told me he cannot get a permit without me paying him first. I did not pay him and he did not help me. I know he would not, even if I pay him.

Afghans who have recently arrived and pay ethnic associations to follow their legal papers reveal other ways that newcomers' illegality is exploited. These narratives are important in revealing the mistrust between irregular migrants and their ethnic counterparts (Suter 2012) and are exemplified in a statement by Ahmed: 'Associations wrote down names a couple months ago. Some of my friends signed up. It does not work anyway. I did not sign up.' What was interesting for the purpose of my research was how migrants sought some level of legitimacy, if not legal status, through these interactions. Mahmut explained that he and a group of Afghan migrants went to the foreigners' department of the police station to register after signing up with the associations and paying the 100 dollar 'donation' required for this process.

> – They sent the list to the police station and we just had to show up. We did not do anything there. We went there as a group. We showed our passport. We were asked when we arrived. We said two years. [Although they were here for one year or less].
> – Why two years?
> – It is to show that we are here, we are interested in staying here, we will not move to Europe. Some other people said four to five years. They made a copy of our passports and let us go.

In this example, migrants seek to gain legitimacy or a kind of immunity from the police through the 'reference' provided by the association and by stating their intention to stay in the country, rather than through acquiring valid papers. Evidence from interviews and literature reveals that neither pro-migrant rights civil society groups, nor migrant associations provide a platform for the political mobilization of irregular migrants. Demands for the regularization of irregular migrants are not on the civil society agenda. Existing ethnic associations rather work as intermediaries to help their clients navigate the changing, shifting legal ground (Parla and Kaşlı 2011). In the absence of political mobilization, most migrants engage in individual legalizing efforts, at times through mobilizing their ethnic identity. Indeed, for some families, it has been possible to acquire permits through their relatives, already settled in Turkey, regardless of their method of entry. Selma and her family, for instance, acquired residence permits because Selma's brother had studied in Turkey, later became a licenced doctor, and eventually a citizen. Others have been eligible for ethnic privilege through the Settlement Law. As mentioned in Chapter 1, Harun's family were also expecting their residence permit at the time of the interview, through their relatives residing in the south-eastern part of the country. This example can be multiplied through further research on other groups of migrants who were admitted earlier, according to the Settlement Law, and granted citizenship. All examples reveal that the right to stay as an immigrant in Turkey operates as an ethnic privilege rather than a right.

In addition to being of Turkish descent, many migrants believe marrying a national is a way to acquire legal status. 'Marriage is the only way to stay here' was a common conviction among migrants interviewed, regardless of their ethnic backgrounds or legal status. Research has already discussed how women from post-Soviet countries get legal status by marrying Turkish nationals (see for instance, Gökmen 2011). Note that in most cases, both ethnic kinship and legal kinship attained through marriage are hypothetical, rather than actual possibilities for obtaining legal status for legalization. Ahmet's reflections, for instance, reveal that there is a lower possibility of getting legal status through ethnicity than through marrying a national:

> I want to go to Afghanistan. I have nothing here but a passport. Before, we could get citizenship; now, it is very difficult, they say. Before, you could get a residence. There are friends who signed up for residence four to five years ago. Now, you can only become a citizen by marrying a Turkish citizen. You marry a Turkish [citizen], you do military service, then you become a citizen. I have a relative like this. He came eight years ago, married two, three years ago, now he is a citizen.

Postponing his ideas of pursuing studies, Chris continued to trade, sending textile goods bought from Merter, a district known for widespread textile production, to his brother so that he could sell them back in Nigeria. He says it is a good business and he is thinking of opening a shop in Nigeria. He might also consider opening a shop in Turkey if he finds a good business partner, who speaks Turkish: 'If I had a Turkish woman, she could help me. I will try sideways, becoming a student, getting married, it is difficult to start a business and get a permit in Turkey.'

This section has revealed the institutional factors behind low levels of mobilization for the rights of irregular migrants in Turkey. Consequently, migrants find few channels to communicate their experiences of widely experienced market violence and the difficulty of accessing fundamental rights without official status. The day-to-day legitimacy partially enjoyed by migrants does not provide political legitimacy for their presence in the urban sphere. In the absence of political mobilization for legal status, migrants turn to individual tactics to secure a legal status.

Conclusion: Turkey as a case of labour market infiltration but limited political mobilization

With respect to irregular migrants' participation in social and economic life and their access to rights and legal status, the Turkish case is characterized by: a somewhat tolerant regime of deportation; the selective participation

of a cheap labour force in the informal economy; limited access to fundamental rights without official status; limited civil society backlash to rights
violations; and a limited ability to claim rights and legal status for migrants
of irregular legal status.

This chapter first suggested that irregular migrants in urban areas have
been largely incentivized to stay quiescent, rather than to protest their lack
of legal status. Practices of deportation and migrants' perceptions of their
deportability have revealed that migrant illegality in the urban sphere has
been conceptualized as 'harmless' within the existing regimes of control.
Given migrants' perception of being tolerated in the urban sphere, their
invisibility becomes strategic. Yet, 'the palpable sense of deportability' (De
Genova 2004: 161), in the absence of available legal structures to legalize
one's status, contributes to the production of a cheap, docile labour force,
as emphasized in the literature on irregular migrants' subordinated status
in the labour market (De Genova 2005; Calavita 2005).

Migrants, even those who allegedly aspire to go to Europe, have found
possibilities to enter the labour market in Turkey as cheap, flexible, and docile
workers. The securitization of the EU borders has arguably increased the
time migrants spend on Turkish soil before transiting to the EU, blurring the
distinction between transit migration and economic labour migration in the
context of Turkey. As a rather unintended outcome of the international production of migrant illegality, those considered transit migrants have become
part of the labour force. Work is available to irregular migrants, but access to
steady jobs remains problematic. Moreover, the conditions of work in terms
of long hours, low wages, and risky conditions (in sectors such as textiles and
construction), reflect the general tenets of the labour market and characteristics of labour-intensive economic growth in Turkey. Hence, incorporation
into the labour force is only possible for young and healthy individuals who
are fit for the kinds of jobs available to migrants. Furthermore, gender and
ethnicity define patterns of inclusion and exclusion in the labour market.

Regarding labour market incorporation, a distinct aspect of the Turkish
case is the interconnectedness between irregularity and asylum regimes.
In practice, there is a thin line between irregular migrants and asylum
seekers, especially for migrants from nationalities that are overrepresented
in asylum applications (İçduygu and Bayraktar Aksel 2012: 8). Both groups
drift between fluid categories, take part in the informal labour market,
and may try to cross to the EU. Meanwhile, the access to certain rights and
services (such as access to residence permits and healthcare) is only possible
by applying for asylum. Throughout the chapter, I have discussed how illegality, labour market infiltration, and asylum regimes at times substitute or

reinforce each other. These interconnections enable migrants' incorporation into either the labour market or bureaucracy, mostly through asylum.

Migrants' incorporation, both through the labour market and through asylum, and the interconnection between the two, reinforce migrants' invisibility and silence in the political sphere. The invisibility itself becomes a way for irregular migrants to present themselves as harmless workers, and hence legitimate members of the society. In the absence of the recognition of basic rights, most NGO activities focus on asylum issues as a legitimate basis for activity. NGOs contribute to the depoliticization of issues related to migrants' human rights when they channel migrants into the asylum track or provide them with humanitarian aid without making explicit political demands on their behalf. Given the lack of allies from civil society or interest from trade unions, as well as a lack of communication among different migrant communities, not to mention the lack of trust within ethnic groups themselves, migrant associations are not powerful or visible enough to make political demands to claim rights and legal status for migrants.

The absence of mobilization for the rights of irregular migrants does not necessarily mean that migrants of irregular status in Turkey are not seeking ways to access rights and legal status. The last section explains how they use existing immigration and citizenship laws to acquire legal status in the absence of communal demands for rights and legal status. At this point, the use of ethnic kinship, envisaged by the 1934 Settlement Law and other clauses in the legislation, help certain ethnic groups of Turkish descent to acquire legal status. Others aspire to achieve legal status and eventually citizenship by marrying a Turkish national or a legal resident.

One can conclude that the lack of strict internal controls and the availability of market opportunities have made it less urgent for those migrants with irregular status who have (semi-) settled in big cities to seek recognition; this is particularly the case in Istanbul. In light of this conclusion, it is essential to reassess the connections between migrants' experiences of deportability, labour market participation, and (dis)incentives for mobilization for their rights. One can question whether migrants of irregular status in Istanbul trade their lack of recognition as rights-bearing subjects for their presence in the labour market being tolerated. In other words, one can also ask whether the market provides a form of de facto exit from the harsh experience of illegality, but ironically becomes one of the factors hindering migrants' associative activities and their political visibility. Chapter 5 reassesses findings from a comparative perspective to reflect on the emerging theoretical discussions.

5 Migrant illegality beyond EU borders

Turkey and Morocco in a comparative perspective

Introduction

This chapter frames migrant illegality and patterns of incorporation in Turkey and Morocco within a comparative perspective. Focusing on the link between migration controls (governance), irregular migrants' participation in society (migrant incorporation), and migrants' access to rights and legal status, the comparison highlights two interlinked questions: First, how does the presence of irregular migrants, despite their lack of legal status, become legitimate within social, economic, and bureaucratic interactions? Second, what underpins the differences in the mechanisms through which migrants gain legitimacy? As promised in the introduction to this book, the comparison aims to explain why certain aspects of migrant illegality and incorporation gain legitimacy over others in particular contexts. Before engaging with the findings of this comparison at a more theoretical level in the Conclusion, this chapter provides preliminary explanations of contrasting mechanisms between the production of day-to-day legitimacy in the absence of a political voice in Turkey and the process of gaining political voice, hence legitimacy, given the very limited forms of daily inclusion in Morocco. In line with the structure followed in Chapters 3 and 4, the discussion in this chapter highlights common and different features of migrant illegality in terms of perceptions of deportability, economic participation, access to rights and to institutions, and in terms of mobilization for legal status that have emerged in both countries.

5.1 Deportations and perceptions of deportability

Interceptions and deportations have become important instruments in the context of the extension of EU migration controls into neighbouring countries. Deportation related practices, such as pushbacks by European border guards, removals to non-EU borders, detention conditions, and denial of access to the asylum procedure, have undermined migrants' human rights on the periphery of the EU. Human rights advocates in both nation-state contexts have criticized such practices. A particularly contested aspect of deportation practices in Morocco has been the removal of migrants apprehended along the Moroccan-Spanish borderland to the Moroccan-Algerian border. Deportations and removal to the border have constituted, until very recently, the main migration control policy of the Moroccan government. Similarly, critiques of the Turkish case have focused on the widespread use of detention as an irregular migration control strategy, the unfavourable conditions of detention centres, and detainees' problematic access to a functioning asylum system.

In both cases, migrants have pushed back from EU borders; those waiting to plan their journeys to Europe have mingled with other groups of migrants that have settled in bigger cities. The blurring of the distinction between 'transit' migrants with alleged aspirations to cross to Europe, asylum seekers, and economic migrants is a common attribute of migrant incorporation, and was seen both in Rabat and in Istanbul, where most of the interviews with migrants in the urban space took place. Policies and practices aimed at stopping irregular border crossings into the EU have rendered migrant groups in both contexts, especially those without legal status, 'deportable' regardless of their aspirations to go to Europe.

Despite similar critiques of Turkey and Morocco's border and deportation policies and practices, there is a striking difference in irregular migrants' perceptions of their deportability from the urban space. Along with juridical status, migrant illegality has a social meaning closely linked to practices on the ground. Notably, in the Moroccan context, mass deportation practices are not limited to border areas. Migrants who are semi-settled in urban areas are also targets of removal. In this sense, deportations are more than a response to EU pressure to keep migrants away from its borders. Security forces have used removal to the border as a primary means of irregular migration control. Parallel with protests about deportation practices along the EU border, stakeholders and undocumented migrants alike have contested the practice of urban raids by the police, articulating that deportation has become part of the daily reality.

It should be noted, however, that despite a perpetual sense of deportability, many migrants in Istanbul expressed feeling at ease with their illegal status in the urban space. Surprisingly, migrants from those nationalities that are most represented in deportation figures, such as Afghans or nationals of African countries, also perceive the police as tolerant of their presence. Similarly, those from African countries who are among the most marginalized groups in the socio-economic sphere because of their physical visibility, recent migration history, and lack of ethnic and linguistic ties have expressed being at ease with security forces. Police raids in urban areas occur occasionally but mainly target drugs, human smuggling, and prostitution related cases. 'The police won't touch you if you are not doing anything illegal' is a common narrative among migrants. Civil society representatives confirm the observation that the police do not systematically inspect urban neighbourhoods and workplaces. Meanwhile, migrants have been subject to random checks, arbitrary practices, and opportunistic abuse. In response, most migrants are acquiescent neighbourhood dwellers and workers, knowing that adopting a less docile approach may endanger tolerance of their illegal status and, in turn, the legitimacy of their presence.

The cases reveal differences in deportability as part of the daily reality in Morocco and as a possibility in Turkey. However, the comparative analysis should not imply that informal arrangements with security forces do not occur in Morocco. Indeed, migrants in the Moroccan context have also occasionally revealed their experiences of being tolerated by security forces despite their illegal or semi-legal status.[1] In both cases, there are groups that enjoy a greater level of tolerance than others. Migrants make conscious attempts to make their presence legitimate. The possession of certain identification papers, even though these are not required legal documents, may be used to avoid trouble with the police and may reduce feelings of deportability. Despite similarities in the way illegality is negotiated at the street level in both contexts, civil society, migrant activists, and non-activist migrants in Morocco have complained about urban raids, police and neighbourhood violence much more than their counterparts in Turkey. This contrast is telling in terms of the connection between migrant deportability and pro-migrant rights mobilization.

The connection between practices of deportability and politicization is worth exploring in both cases. Peutz and De Genova (2010: 19) have suggested that deportability does not necessarily make migrants passive; on the contrary, it may also mobilize them towards collective action. However, the connection between deportability and collective action cannot be taken for granted. The situation that has given rise to widespread mobilization in the Moroccan case should be contextualized within the broader policy context. First, as explained in Chapter 2, the ongoing state-led politicization has depicted irregular migration as a security threat since the early 2000s. In relation to this criminalization of irregular migration by law, mass deportations are directly used to curtail irregular migration on Moroccan soil. Criticism of police and border violence by pro-migrants' rights

---

1   According to the GADEM report, tolerance by security forces is more common in Southern Morocco where most migrants overstay their visa and migrants have more opportunities in the labour market (GADEM et al. 2014).

group, including migrants' associations, have emerged in response to this particular top-down politicization process.

Activists in Turkey have also contested deportation and detention practices, albeit less intensively. One can rightfully ask why there has been a limited mobilization in Turkey. On the one hand, Turkey seems to display lower degrees of state-level politicization of the presence of irregular migrants in the country. Irregular migration policies are seen within the technical aspects of the EU accession process. The rather low level of politicization of the issue is coupled with law enforcement officers turning a blind eye to the presence of irregular migrants, especially to those participating in the informal urban labour market.

5.2   Socio-economic participation and daily legitimacy

The connection between migrant deportability and migrant mobilization can be better comprehended by looking at the functioning of migrant illegality in social and economic life (Coutin 2003; De Genova 2002; Calavita 2005; Willen 2007a). As widely discussed in irregular migration literature, migrant illegality and deportability typically result in subordinated forms of inclusion of irregular migrants into society rather than absolute exclusion (Chauvin and Garcés-Mascareñas 2012). This inclusion may happen as an unintended effect of unprecedented irregular human mobility, or as part of the specific political economic agenda by leaving the door open to the arrival of irregular migrants. Regardless of intent, their participation in social and economic spheres may legitimize irregular migrants' presence in a given territory, despite their perpetual perceptions and experiences of deportability.

Irregular migrants settling in disadvantaged areas of the city and working in the informal economy in a sporadic fashion is a general and common characteristic of subordinated incorporation in Rabat and Istanbul. This informal incorporation has given rise to different forms of violence and marginalization in both contexts. Exploring different degrees of informal economic activities pursued by migrants in both contexts enables us to question how experiences of illegality in economic life have implications for the political presence and legalization strategies of migrants.

Housing is an initial step to migrants' economic participation in the receiving society. In Istanbul as well as in Rabat, migrants are concentrated in poorer areas of the city, and they are initially accommodated by informal reception mechanisms such as informal employment and real estate agencies, relatives, and co-ethnics who are already settled in the area. In Istanbul, proximity to work opportunities is another factor that influences migrant settlement in the urban space. Poor living conditions, such as overcrowded rooms and flats that are mostly, but not exclusively, shared with co-ethnics and accommodation lacking proper sanitary facilities are significant housing issues. Informal contracts with property owners, overpriced rents (in comparison to what locals pay), and the quality of housing offered reveal crucial aspects of the economization of the presence of irregular migrants in both contexts. In Istanbul, the arrival of Syrians has reportedly increased rents even further. Despite this common economization through housing, there are differences in the degree to which the migrant labour force has become part of the informal economy in both contexts.

In comparison to Rabat, the structure of the labour market and the scale of informal economic activities in Istanbul are more suitable for accommodating irregular migrants. Hence, there is widespread migrant participation in the already established informal economy, characterized by an unregistered workforce in the textiles, construction, domestic work, tourism, and service sectors, which has been a distinctive aspect of their informal incorporation experience. Informal employment agencies and ethnic networks facilitate migrants' inclusion into the labour market. In line with previous research, findings have underscored that the informal labour market in Istanbul does not necessarily distinguish between migrants with and without legal status. Similarly, not only migrants who arrive in Turkey in search of employment opportunities, but also those with aspirations to cross to the EU and/or fleeing from conflict have found positions, albeit precarious, in the labour market. Despite the recent history and scale of irregular migration from diverse locations into Turkey and the temporary character of migrant settlements, it may be possible to identify ethnic enclaves within the urban economy. Sectors such as domestic care have been widely employing migrant women from former Soviet Republics such as Moldova, Georgia, and Turkic Republics. Women from these countries are also known to engage in the textile trade. There are large numbers of Afghans in the leather and textile industries, whereas Western Africans are rather associated with street vending and commodities, particularly textiles and trade.

Migrant deportability has also given rise to a young, exploitable migrant labour force in Morocco. However, the analysis of migrant economic participation in the case of Rabat reveals that the economic gain of irregular migrants in the housing market does not necessarily translate into migrant incorporation in the informal labour market. Unemployment or working sporadically (a few days per month) seems more common among migrants in Rabat than their counterparts in Istanbul. In the context of widespread exploitation and under-payment, wages fall short of covering basic expenses, let alone savings to finance a further journey or to send back home and this is another factor in their exclusion from the labour market. In construction, for instance, apparently the most suitable sector for young male migrants, migrants complain that their daily wages are half those paid to locals (see also Pickerill 2011).[2] This situation pushes migrants away from intensive work in the waged labour market towards daily income generating activities (street vending, hair styling, etc.). Some also engage in precarious activities such as begging or prostitution. Only a minority has stable jobs, for example West African (mostly Senegalese) or Filipina women working for upper-middle-class Moroccan or expat homes as domestic workers or looking after children, and a small group of educated migrants (some are former students from Western and sub-Saharan African countries) work in call centres. However, even migrants with a legal entry and educational qualifications have found it difficult to find a steady job and to legalize themselves through employment contracts.

Differences in scales of economic activities between Rabat and Istanbul may explain the striking difference in terms of access to labour market opportunities. In terms of GDP, Turkey's economy is nearly eight times bigger than Morocco's. As the main economic hub with a large informal economy, Istanbul attracts migrants seeking economic opportunities as well as those with explicit aspirations to go to Europe. Moreover, Istanbul's urban economy has experienced a significant transition, including the expansion of, among others, the construction and textile sectors. Since the 1980s, economic growth has been dependent on lowering labour costs. This was also a time when labour unions were weaker than in previous periods (Çelik 2013) and, indeed, they are now virtually non-existent in certain sectors where sub-contracting and informal labour are widespread (Toksöz, Erdoğdu, and Kaşka 2012: 23). The IOM report on irregular labour migration in Turkey indicates that the net wages of irregular migrants are not necessarily lower than the wages of locals (Toksöz, Erdoğdu and Kaşka, 2012: 23). However, as employers avoid taxes and social security expenses and demand longer working hours, hiring irregular migrants significantly decreases labour costs and thus remains an attractive alternative to local employees. Initial reports suggest that the arrival of Syrians has actually

2   As of 2012, wages for daily jobs in construction, for instance, were reportedly between 55 MAD and 80 MAD.

lowered wages in certain regions, particularly in sectors such as seasonal agricultural work, textiles, and construction (Kavak 2016).

Most striking in the analysis is not the absolute exclusion from income generating activities, but that migrants in Rabat are more inclined to express their marginalization because of the lack of labour market opportunities. In comparison, despite their positive views on labour market opportunities, migrants in Istanbul are more likely to complain about exploitation in the labour market. Temporary work arrangements suitable for younger migrants, high turnover rates, moving from one workplace to another, and from one sector to another, are common among migrant workers. In this sense, similar to what is happening in Morocco, the labour market in Turkey offers few possibilities for social mobility or for legalization through work contracts.

Despite differences in the intensity of available work opportunities and of labour market participation experiences in these two settings, the research has revealed that labour market incorporation is a selective process. Work is available for young, able bodies who can endure hard labour conditions. Pregnant women and women with younger children are the most marginalized groups in the labour market. In the absence of labour market possibilities, migrants considered as vulnerable rely on support from humanitarian organizations. In this sense, the legitimacy of their stay does not stem from their contribution to the economy, but rather their vulnerability. Begging and marginal ways of generating income are widespread especially among these most marginalized communities. For instance, in the case of Morocco, Nigerian women with babies are associated with begging and sex work. Recently, Syrian women and children have suffered the same stigma in the Turkish context.

One implication of migrant labour concentration in Istanbul has been the blurring of the distinction between registered asylum seekers, people with asylum claims who are not registered with the authorities, and unregistered (so-called) economic migrants. The asylum system in Turkey does not provide prospects for permanent legal status in Turkey or for resettlement to a third country in the near future. Consequently, asylum seekers may breach the asylum regulations that require them to reside in their assigned province and instead come to Istanbul to generate income in the informal economy. Some migrants with potential asylum claims do not even apply in the first place, knowing they would need to leave Istanbul and all the economic opportunities the city offers. The implications of the specific asylum regime are a distinctive part of the production of migrant illegality in the labour market in Turkey. As already discussed in Chapter 2, the new law, the LFIP, aims to reinforce the distinction between asylum seekers entitled to legal status and irregular migrants. Conversely, the distinction between legal and policy categories such as asylum seeker vs. irregular migrant, or transit vs. economic migrant, remains blurred in practice. The mismatch between legal requirements and the labour market has created a situation in which migrants, especially potential asylum seekers are forced to choose between a less precarious but 'liminal legal status' (namely asylum) (Menjívar 2006), which limits mobility within the country, and the precarious labour market opportunities concentrated in big cities without any legal status. In this sense, Turkey has been an example of how irregular migration regimes in interaction with asylum regimes produce deportable, flexible, and cheap labour.

As discussed above, migrant illegality in the urban sphere is tolerated as long as the authorities, mostly street-level officials, are convinced that migrants' economic endeavours are not linked to crime-related activities such as human smuggling, prostitution, drug dealing, etc. The character of the widespread but precarious (temporary and low paid) employment reinforces the image of a docile migrant worker. Labour market participation provides a degree of protection from deportation, if not from police intervention and occasional harassment. Meanwhile, irregular migrants find themselves in a vulnerable situation in terms of access to fundamental rights and legal status. As explained in the section 4.4, the particular (de) politicization of immigration-related issues and the weak pro-immigrants' rights movement in Turkey have been factors that have contributed to the silencing of labour market violence in Turkey. As a consequence, another indirect implication of labour market incorporation in Turkey has been that migrants find no channels for raising a political voice to improve their labour market situation. This lack of mobilization needs to be contextualized within the general silencing of labour-related issues in Turkey.

In contrast, in Morocco, social and economic marginalization, coupled with strict migration controls, have characterized the 'origins of the suffering of irregular migrants,' hence 'the objective context' leading to mobilization by civil society, but also by irregular migrants themselves (Chimienti 2011: 1340). Chapter 3 explained that the street violence against migrants from sub-Saharan countries and the backlash against such xenophobia are widespread. Experiences of exclusion are not only a result of marginalization in the labour market, but also of racist street violence. In Morocco, both exclusion from the labour market and widespread racial aggression are prevalent in migrants' daily perceptions of their illegality and they are widely expressed by migrant groups and pro-migrant rights actors.

In comparison with the situation in Morocco, street violence has not been a central theme in the narratives of migrants and stakeholders interviewed in Istanbul. Again, one should be careful not to imply that there is no street-level violence, racism, or discrimination against migrants in Turkey. Even though cases of aggression exist, they are not central to migrant experiences of illegality, unlike the case of Morocco. There have been few accounts of aggression against migrants in both the media and extant literature. Instead, what is central to their physical/bodily experience of illegality is the exploitation stemming from labour market conditions. Meanwhile, it is worth acknowledging that migrants' participation in the labour market has given a degree of legitimacy to their presence. However, this form of daily legitimacy does not necessarily translate into political activism, despite widespread forms of labour market violence. As further elaborated below, factors underpinning a lack of mobilization include the absence of politicization, which stigmatizes migrants at the policy level, of active repression mechanisms, plus, irregular migrants' lack of access to pro-migrant rights channels, and the weakness of the asylum process.

5.3 Access to rights through institutions and the role of 'street-level advocacy'

Subordinate inclusion in the informal/secondary labour market is not the only mechanism legitimizing the presence of irregular migrants in society. The formal political authority may also indirectly recognize the presence of irregular migrants by enabling their access to fundamental rights despite their lack of legal status. Along with economic incorporation, mechanisms of bureaucratic and political incorporation influence migrants' access to rights and to legal status and may provide de facto recognition of their 'illegal' but legitimate presence in a nation-state territory.

Migrants and pro-migrant rights actors negotiate illegality within the sphere of fundamental rights, beyond market relations. Migrant illegality can be reversed through access to fundamental rights by enabling access to state institutions providing services, despite their exclusion from the legal sphere. Migrants usually need support from pro-migrant actors to surmount the bureaucratic mechanisms that deny them their fundamental rights, even when the law recognizes these rights. The mechanisms that give migrants access to rights differ from one context to another.

In both Turkey and Morocco, migrants' needs for healthcare and education are at stake as an indirect result of enduring migrant illegality and because they are stranded. In urban areas, health problems stemming from poor living standards and lack of hygiene increase migrants' needs for healthcare. Reports on Morocco underscore that such needs are even more urgent in informal camps along the borders between Morocco-Algeria and Morocco-Spain, where living conditions are even harsher and physical injuries are common because of clashes with security forces. Children have been a minority of immigrant groups in both contexts; however, their numbers and visibility are increasing, as more families have settled in urban areas. Some families come with children, while others have children along the journey or during their stay in Turkey or Morocco. Thus, minors' access to education has become a legitimate concern, regardless of parents' aspirations to cross into the EU or to settle in Turkey or in Morocco. Minors' access to public education also has symbolic importance within discussions of the integration of immigrants and of membership in both nation-state contexts.

Findings reveal that in both Turkey and Morocco, irregular migrants' access to fundamental rights is problematic at the legislative level, i.e. at the level of recognition, as well as at the level of enforcement. In terms of migrants' access to rights, both states initially denied responsibility towards immigrants on their soil. Improvements in respective legal frameworks in both countries have recognized irregular migrants' access to fundamental rights, albeit in a very limited fashion. According to national legislation enacted in Morocco in 2003, irregular migrants should have access to free primary consultations as a public health concern and within the context of preventing epidemics. The asylum regulation in Turkey initially did not include provisions to cover the healthcare expenses of asylum seekers and refugees. Since 2008, free or subsidised public healthcare is possible with an official status (e.g. asylum seekers, refugees, stateless people, foreigners with a residence permit) depending on one's declared income. Asylum applicants were only included in this scheme in 2013. Thus, irregular migrants have not even been part of the discussion of access to healthcare. Based on international conventions ratified by governments and on constitutional principles in both contexts, children have the right to public education regardless of legal status.

At the level of enforcement, bureaucratic obstacles hinder migrants access to these rights. The access to basic services, even with an official status or when the law recognizes these rights, is not straightforward. In both cases, civil society interventions have been instrumental in making laws work in practice. NGOs also play a crucial role in day-to-day advocacy by negotiating bureaucratic obstacles and, to a certain extent, in surmounting migrants' exclusion from the realm of rights.

The general observation is that, in both contexts, migrants without legal papers (such as a residence permit, asylum application, or at least a passport with a legal entry) are not admitted to public hospitals, and, at times, even to emergency rooms. In Turkey, when migrants are admitted to hospitals, they might be asked to pay higher fees, so-called tourist fees, if they are not registered within the general health insurance system. It is common for migrants to have to cover their own health expenses in Turkey, but also in Morocco, with respect to secondary level treatment such as diagnostic analysis or hospitalization. In this sense, there is a concrete problem in terms of access. The general (mal)functioning of national healthcare systems reinforces migrants' bureaucratic exclusion from public health care.

Given the problems regarding the functioning of laws, interventions by civil society are aimed at meeting urgent humanitarian needs of migrant communities. NGOs play a complementary role in the sense that they provide basic healthcare and informal education when access to public service is not possible. Due to the urgency of the situation, international funding for such activities has been available. Project-based civil society activities prioritize vulnerable groups such as pregnant women and unaccompanied migrants as well as problematic areas such as borderlands. Such projects have surely made a difference in meeting migrants' urgent needs, in reaching vulnerable groups, and in appeasing their sense of exclusion from public institutions. However, they are limited in their scope. Rather than formally recognizing rights, these practices indicate a general trend that sees welfare state services channelled towards civil society, generally with limited resources.

While problematizing the role of NGOs in providing services that, conventionally, are the responsibility of states, I suggest these practices play a twofold role in terms of advocacy. First, besides being direct providers of humanitarian aid and services, an emerging civil society, in both Turkey and Morocco, has actively engaged in the 'cumulative creation of law' (Coutin 1998). Second, the cumulative creation of law may lead to informal and eventually to formal access to rights, hence to the formal recognition of the presence and legitimate rights of migrants without legal status. Civil society actors have worked towards the enforcement of rights that are recognized but which are not properly implemented at the institutional level. To this end, they have engaged in daily negotiations with street-level bureaucrats, such as school principals and chief physicians in hospital departments that admit migrants. What I call 'street-level advocacy' may, at times, turn into informal agreements between civil society and state or private institutions. One common response to the problems of implementation regarding access to healthcare has been to make informal agreements with service providers. For instance, NGOs transfer those in need of medical care to hospitals they work together with and that are more familiar with receiving immigrants with no proper identity papers. This practice ultimately aims at ensuring migrant access to these institutions without civil society intervention. However, 'autonomization of patients' by MSF (author interview, Rabat, April 2012), namely the idea that migrants, regardless of legal status, can reach these services by themselves, is not likely to happen in either Morocco or Turkey. Irregular migrants continue to rely on civil society connections to get access to public hospitals. NGOs have to renegotiate their informal arrangements on a daily basis because of the non-standardized institutional behaviour and the changing legal framework.

What we might call 'cumulative creation of law' provides possible openings to universal access and, in turn, recognition. Yet, it may also reveal differences between migrants with no legal status and asylum seekers, legitimatizing the access of asylum seekers to fundamental rights at the expense of legitimate access for all. Because of the restricted legal framework in Turkey, migrants' access to rights is only possible with an official status. The status of asylum applicant is the only official status that irregular migrants, especially those without legal entry and those who fit the profile of asylum seekers, can acquire in Turkey. In other words, the restricted legal framework makes asylum the only option for achieving legal status, hence access to healthcare and education. In urgent cases, such as injuries or pregnancy, NGOs fast track migrants into the asylum process using their connections to UNHCR and hospitals, as it is the only way to get access to public healthcare. Also, despite universal access embraced in the legal framework, only children of asylum seekers residing in their assigned satellite cities can get formal access to schooling. Others may be accepted as guest students depending on informal arrangements between school principals and civil society. The prerequisite of an official status forces migrants to choose between labour market opportunities in big cities and the right to healthcare and schooling in satellite cities where economic opportunities are scarce. In the absence of legal status, irregular migrants, left to their own devices, refrain from seeking healthcare unless absolutely necessary. In emergencies, they resort to their community networks or the private market. In this sense, access to healthcare is no longer a form of bureaucratic incorporation but becomes another form of the economization of irregular migration. Evidently, the market option is only possible if migrants can afford it.

Informal negotiations by civil society may lead to formal changes towards more inclusive practices that enable migrants to acquire legal access to rights. In such endeavours, claiming rights for asylum seekers may provide an opening for all migrants regardless of legal status. In contrast with the Turkish case, the case of schooling in Morocco reveals how a semi-formal arrangement between UNHCR and the provincial public education directorate concerning the children of recognized refugees and asylum seekers has been used to enrol all migrant children regardless of legal status. Access to public education has been possible through bureaucratic camouflage, mingling children of irregular migrants with those of asylum seekers. What is more interesting is that, following demands by civil society and by migrants themselves, a new regulation on this issue has enabled children's access to public schools regardless of their parents' legal status. The related regulation making requirements for school registration more flexible is one of the concrete steps in a new policy initiative in Morocco launched in November 2013. Meanwhile, a level of self-exclusion by migrants themselves has been visible. For instance, Christian migrants refuse to send their children to public schools where Islamic education is an integral part of the curriculum. Parents' aspirations to further their journeys to Europe constitute another reason for excluding their children from accessing their rights. The impact of this change on integration is yet to be seen. However, reformist steps towards recognizing the right to education have increased recognition of migrants' fundamental rights regardless of legal status.

From the comparative perspective, legislation coupled with civil society interventions in Turkey reverse migrant illegality by turning irregular migrants into asylum seekers. This is an example of what Coutin (2003) describes as 'legalizing moves'. At the same time, this practice reinforces the distinction between legitimate grounds for asylum and the illegitimate presence of irregular migrants subject to deportation and very limited access to rights, already explicit in the new LFIP legislation. In other words, NGOs subscribe to the limitations of the existing legislation rather than pushing for more inclusive practices for formal recognition of migrants' fundamental rights regardless of legal status. The mechanisms for accessing fundamental rights pull migrants into the system by turning them, first, into asylum seekers, and second, into clients with very marginal benefits within the welfare system. In other words, redistribution in terms of access to certain rights is tied to a very particular form of recognition as an asylum seeker. Conversely, in the case of Morocco, NGOs have mainly worked towards the enjoyment of fundamental rights by all migrants regardless of legal status. Such inclusive attitudes have, arguably, reinforced migrant mobilization and their quest for legal status through collective action. In both contexts, there is an opening in terms of migrants' access to fundamental rights, but

I would argue that these openings have carved out different trajectories for political action.

5.4   Reversing illegality

In light of the previous comparative analysis of practices and experiences of deportability, of labour market participation, and of access to fundamental rights, this section discusses the political legitimacy of mobilization for the rights of irregular migrants. It questions the circumstances under which irregular migrants' rights have or have not constituted a legitimate sphere for civil society advocacy. How have irregular migrants, in return, become part of the mobilization process and actively claimed rights and legal status in the context of Morocco but not in Turkey? Mapping civil society in a comparative perspective helps us to account for differences in the mechanisms of access to fundamental rights discussed in the previous section, as well as for differences in advocacy. A comparison of the activities and priorities of pro-migrant rights civil society explains differences in migrants' strategies for achieving legal status.

Mobilization for the rights of irregular migrants

Both in Morocco and Turkey, civil society interest in immigration issues emerged in the post-2005 period, as an indirect outcome of the externalization of EU migration policies, but also as a response to human rights violations and the urgent humanitarian needs of migrants. Violence by security forces in urban areas and at the borderlands around the time of the events in Ceuta and Melilla, in 2005, resulted in a turning point, and led to the emergence of civil society actors interested in the question of incoming migration to Morocco. In Turkey, the EU accession process provoked legal changes on border, asylum, and irregular migration issues, which, in turn, triggered civil society activities. IOM and UNHCR entered the field of immigration in both counties in the post-2000 period. These organizations have been led the way for local civil society. In both contexts, several local organizations have become service providers for UNHCR. For other advocacy and humanitarian organizations, international funding by the EU as well as by other international funders has been available. Despite the similar political contexts in terms of strict regulations on NGOs, civil society working on immigration and asylum issues, gradually expanded its expertise and activities on the subject. In this sense, analyses should take into account that externalization of border controls also leads to externalization of humanitarian intervention and of expertise.

Both contexts display a similar mapping of civil society actors. On the one side, there are international, church-based, national, and more local organizations primarily concerned with providing humanitarian aid to migrants. On the other side, there are human rights organizations engaging in advocacy and providing legal aid. One significant difference in the Moroccan context has been the institutionalization of irregular migrant associations. In fact, migrant associations, which already existed as mutual aid societies in the rural and urban areas, have gained political visibility in the post-2005 period.

Both countries also display commonalities regarding civil society activities for migrants and in civil society's relations to state organizations. Civil society ensures migrants' access to fundamental rights despite relying on different sources to ensure irregular migrants' legitimate access to fundamental rights, as discussed in the previous section. Civil society has played an important watchdog role in revealing rights violations against migrants. The advocacy activities embraced by civil society led to tense relations with the state, but these later evolved into limited forms of cooperation. Note that such cooperation, generally initiated by the state, has been welcomed, but, at the same time, has been received with suspicion by civil society.

As explained above, the categorical separation of asylum seekers and irregular migrants, reinforced by most NGO practices in Turkey, has direct implications for NGO advocacy. Advocacy activities by civil society in Turkey have emphasized problems related to asylum seekers. Human rights organizations have reported on asylum policies, the malfunctioning of asylum system, and the conditions of detention. Through lawyers specialized on migration issues, NGOs have asked the ECtHR to take interim measures to stop unlawful detentions and deportations. In this sense, the main references and sources of legitimacy for civil society's critique of the state have been the ECHR and EU conditionality as well as the 1951 Convention, although Turkey does not grant refugee status to applicants from non-European countries. ECtHR decisions against Turkey concerning the treatment of detainees have worked as transnational advocacy mechanisms initiated by national civil society and have, arguably, accelerated the drafting of the law on Foreigners and International Protection (LFIP) in Turkey. The process of law-making and the establishment of the Bureau of Migration and Asylum under the Ministry of Interior has provided the basis for rapprochement between civil society and state institutions since 2008.

In the case of Morocco, irregular migration and the right of irregular migrants to stay in the country have been at the centre of NGO activities. Interception and deportations of groups such as minors, pregnant women, and asylum seekers, who are protected within the existing legislation have been subject to critique by national and international civil society actors, including migrant associations themselves. Similarly, mass deportations (most of the time without access to judicial review or appeal) have provided grounds for contestation by civil society, including migrant associations. Civil society has actively worked to raise awareness of xenophobic violence. Furthermore, the social and economic rights of irregular migrants have been a central focus of civil society activities. In this context, the 1990 Convention on the Protection of the Rights of All Migrant Workers and Members of Their Families regardless of legal status has been the main reference. The unionization of migrant workers has also become part of the mobilization process in Morocco. Already existing demands for regularization of irregular migrants have gained a formal character through the formation of a union for migrants under a nationwide and sector-wide worker's union in the country. Below, I detail a distinctive aspect of the Moroccan case, in terms of migrant associations forming alliances with Moroccan associations to gain political recognition through the mobilization process. Despite ongoing tension after the reform initiative in Morocco, the state institutions have held regular meetings with civil society. The legal recognition of informal migrant associations has been on the agenda during these meetings.

Unlike in Morocco, civil society interest in Turkey for issues pertaining to irregular migration has not gone beyond humanitarian support and has not evolved into a radical discourse claiming rights and legal status for all regardless of legal status. In contrast with the case of Morocco, Turkish unions have not developed an interest in including irregular migrants into their membership base, despite an increase in labour migration into Turkey. The general lack of interest on the part of unions in Turkey is partially due to their absence in the informal sector. Unions consider the informal market as a structural problem that can only be fixed through formalization; rather than as an integral character of the labour market that should be incorporated.

In the absence of civil society embracing irregular migration issues, such demands for regularization are not on the agenda of NGOs in Turkey. An exceptional regulation introduced in the summer of 2012 has enabled migrants overstaying their visa to regularize their legal situation. However, this legal intervention was a top-down state initiative, rather than a response to grass-roots demands. Migrants had to pay high fees for the time they overstayed their visa in return for a short-term (six months) non-renewable residence permit. Beneficiaries of this regulation, who could not secure their longer-term residence permits, have fallen into irregularity at the end of the period they were regularized. Therefore, they perceived this legal move not as recognition of their right to stay in Turkey, but as another form of the economization of irregular migration.

Regarding the divide between asylum and irregular migration among civil society actors in Turkey and with the absence of such a divide in Morocco, there are exceptions in both contexts. Few civil society activities in Morocco have been exclusively limited to asylum seekers, whereas some organizations in Turkey have embraced more inclusive practices and demands. There are few network-type organizations that explicitly refute the divide between asylum seekers and refugees. Recent protests on street violence and labour market violence towards immigrants, as detailed in Chapter 5, demonstrate the unprecedented involvement of actors such as groups working on labour issues or feminist groups. Although these groups do not necessarily work on immigration issues, they have gradually developed an interest in the vulnerable conditions of irregular migrants in the labour market. Nonetheless, such initiatives have had a limited sphere of activity and influence compared to formally established NGOs.

Low levels of politicization of irregular migration by the state, hence the lack of public opinion formation processes on irregular migration and asylum issues, may explain this belated interest by civil society actors (Tolay 2012). Although this situation has started to change with the arrival of Syrian refugees, issues related to non-Syrian refugees and irregular migrants are silenced. It is worth elaborating the general pattern of how and why the rights of irregular migrants have the focus of advocacy in the context of Morocco while they are sidelined in the context of Turkey.

A lower degree of politicization of irregular migration by the state shapes civil society activities in Turkey. Differences in asylum tolls and the emergence of the question of irregular migration in the two contexts underpin the difference in UNHCR's impact. As explained in Chapter 2, Turkey receives a much higher number of asylum seekers compared to Morocco. Moreover, in Turkey, international migration became part of the political agenda as an asylum issue with the arrival of asylum seekers fleeing Iran in the aftermath of the Islamic Revolution, around the same time as the closure of European borders. These flows were coupled with asylum seekers fleeing a post-military coup Turkey. In this sense, the discussions on refugees dominated public opinion since the 1990s, even before discussions on transit migration (Hess 2012: 431). As a consequence, UNHCR's impact on civil society in Turkey has arguably reinforced the dichotomy between refugees 'who are constructed as being in need of protection and whose cross-border movements are recognized as legitimate, and "illegal" migrants, whose movements' legitimacy is denied' (Scheel and Ratfisch 2014: 928).

Civil society organizations that are actively working on asylum issues recognize the existence of irregular migrants in need of civil society support. In this sense, my findings differ from those of Şenses (2012: 208), who argues that civil society actors do not have an informed opinion about the question of irregular migration and that they do not have a clear pro-migrant attitude. Meanwhile, the advocacy language embraced by NGOs in Turkey has prioritized the storyline of asylum seekers rather than referring to migrants in general. Partly because of arbitrary tolerance towards irregular migrants, explained in Section 5.1, civil society critiques have focused on the conditions of detention and access to asylum after apprehension, rather than the daily forms of abuse faced by irregular migrants in social and economic life. Given their limited resources, they consider it more legitimate to negotiate the rights of asylum seekers who are, on paper, under international protection.

UNHCR in Morocco has endeavoured to propagate its own discourse among civil society actors in Morocco, on 'mixed flows' and on the necessity to distinguish between those in need of genuine international protection and economic migrants. However, most NGOs have resisted such clear-cut definitions (Alioua, 2011: 475-6). Asylum seekers correspond with a small portion of the migrant population in Morocco. At the level of state practices, there is no legal distinction between refugees and irregular migrants. In terms of objective conditions, refugees and irregular migrants are subject to similar deportation and violations. Therefore, it makes no sense for civil society actors to focus on asylum issues, as UNHCR has encouraged them to do.

Another important factor making irregular migrants the focus of civil society attention in Morocco has been the political framing of the issue. Trans-Saharan transit migration coupled with clandestine movements into Spain was highly politicized and illegalized as a security issue (Andersson 2014). It has mostly been conceptualized in Moroccan policy discussion as an economic migration issue, rather than an issue of asylum. All these factors, namely the profile of migrants, the experience of harsh controls by refugees and migrants alike, and the framing of irregular migration as a security problem, have contributed to the emergence of irregular migration as a legitimate sphere for policy interventions by civil society. In comparison to Turkey, the level of mobilization for the rights of irregular migrants and the role played by migrants' associations in these networks are much more visible in Morocco.

Migrant mobilization for legal status

In response to different types of exclusionary practices and human rights violations, migrants in Morocco formed solidarity associations. The Ceuta and Melilla events, forming the 'transition moments in migrants' engagements', in the words of Però and Solomos' (2009:11), pushed existing solidarity networks to become political organizations. In addition to migrants being pushed towards collective action by the politicizing effects of experiences of illegality, subordinate incorporation, and being stranded; migrant mobilization in Morocco has fostered support for such action by Moroccan and international NGOs. As sub-Saharan migrant associations became more established and collaborated more frequently with Moroccan and transnational civil society actors, they amplified their visibility and their demands for the fundamental rights of migrants, the regularization of undocumented migrants, and the formal recognition of their associations. While the mobilized groups also include asylum seekers and recognized refugees, coordinated demands for regularization, i.e. the right to stay on Moroccan soil, proves that, in essence, mobilization has mainly been a movement of and for *sans-papiers* ('those without documents'). Migrants have been actively taking part in street demonstrations, advocacy activities, and eventually in policymaking. In this sense, migrant mobilization in Morocco has been a process of gaining political legitimacy as opposed to continuing physical, social, and economic exclusion of sub-Saharan migrants in Morocco.

Migrants in irregular situations in Turkey do not display a similar level of mobilization to claim rights and legal status as their counterparts in Morocco. The absence of collective action by irregular migrants is surprising given the similar experiences of being stranded as a result of the difficulties of crossing into Europe and the experiences of marginalization in social and economic life. Common experiences of illegality, as well as African identity and a shared knowledge of French, have enabled the mobilization of migrants in Morocco. It is fair to suggest that such identity-based mobilization is less likely to happen in Turkey, where migrant profiles are more diverse, in terms of an ethnic and linguistic background, and a legal distinction between asylum and irregular migrants has been more clear-cut. However, even within communities sharing language and/or nationality, political mobilization has been limited. On the one hand, irregular migrants, especially those of African origin, have formed solidarity communities and informal networks based on national background, but they tend to be short-lived and invisible in the public sphere (Suter 2013: 205-7). On the other hand, ethnicity-based formal associations established by previous immigrants who later gained citizenship have, to a degree, included newly arriving co-ethnics as informal members. However, the activities of such associations do not entail lobbying for the rights of irregular migrants. Social hierarchies between those with legal status and newly arrived undocumented migrants have inhibited irregular representation issues in these associations. In this sense, irregular migrants have not been able to use existing ethnic associations as a platform for raising a political voice.

As I have already suggested, the lack of migrant mobilization in Turkey is the result of a process of depoliticization produced at the intersection of socio-economic, institutional, and legal fields. The factors that have pushed for and enabled migrant mobilization in Morocco have been absent in Turkey. The controls and deportation practices leading to the politiciza-tion of existing migrant solidarity communities in Morocco were not as harsh in Turkey. Rather, irregular migrants in the urban context in Istanbul experience day-to-day legitimacy, mainly through their labour market participation. One can ask whether higher levels of economic incorporation replace irregular migrants' quests for recognition and political legitimacy. In other words, it is important to look at the precise conditions under which we can talk about a trade-off between political activism leading to political legitimacy and day-to-day legitimacy coupled with forms of arbitrary toleration. This is where institutional factors come into play. Unlike Morocco, there is no civil society support for irregular migrant mobilization in Turkey. CSOs conceptualize asylum seekers and, to a lesser extent, irregular migrants, mainly as beneficiaries of their services, rather than as political actors. Public demonstrations on the subject are rare, and migrants in Turkey have only been participated in such contestations of state practices on exceptional occasions.[3]

Direct references to the situation of Moroccan migrants in Europe has provided a legitimate basis for explaining why Moroccan civil society is interested in migrants in irregular situations in Morocco. Such references have also enabled the transnationalization and expansion of the movement within and beyond the country. Emigrant associations such as ATMF have demonstrated an interest in this issue, together with state institutions,

---

3   The hunger strike by Afghan refugees in Ankara was one of the exceptional protests where refugees were at the frontline. However, the target of the protest was UNHCR, which had suspended asylum applications from Afghan nationals, rather than the Turkish state.

including semi-public institutions dealing with issues related to Moroccans abroad, such as CCME. Arguably, the *sans-papiers* movement in France has influenced pro-migrants' rights mobilization in Morocco. The linkages are salient in terms of repertoires of rights claims and of transnational actors involved in the struggle. In other words, some activists supporting migrant mobilization in Morocco have been part of the movement in France. The French language has facilitated communication and forged alliances in this regard. In the absence of a process of the politicization of migrant mobilization, such references enabling the support of transnational actors have also been absent in Turkey. One reason that Turkey's emigration history is almost never raised in the discussion of immigration-related issues is that labour outmigration from Turkey has been seen as *passé*, rather than an ongoing reality. Again, given the prevalence of asylum issues, stakeholders may find it hard to build connections between the vulnerable situation of 'incoming refugees' and state-coordinated economic outmigration to Europe in the 1960s and 1970s.

The empirical analysis acknowledges that a minority of migrants are mobilized within associations in Morocco. I also refrain from implying that there is absolutely no mobilization for the rights of irregular migrants in Turkey. However, the general trend of lower level mobilization for the rights of irregular migrants in Turkey is striking. Rather than having a political/ activist migrant identity, revealed in a number of cases in Morocco, most migrants interviewed in Istanbul linked their prospects for legal status to individual legalizing efforts. For those who fit the profile of asylum seekers, acquiring asylum status is one way of obtaining legal status, although asylum status does not automatically lead to access to rights, as explained in section 4.3. Most migrants, also aware of the (mal)functioning of the asylum system, do not consider asylum as route to legalization in Turkey. Marriage with a Turkish national and being from Turkish descent are the most common ways for migrants to acquire legal status. Rather than the recognition of migrants as rights-bearing subjects through institutional means, the legislation enables legal incorporation for some of these migrants as ethnic kin or 'legal kin' when they marry a national.

Conclusion

This chapter has discussed the empirical findings of Chapters 3 and 4 and explored patterns to generate a hypothesis. It questioned how irregular migrants rendered illegal and 'rightless' legitimize their right to stay in the territory. Answering this question, the comparison reveals, in particular, how daily experiences of illegality and mechanisms of irregular migrant incorporation have given rise to different ways for irregular migrants to gain legitimacy. While day-to-day legitimacy has been the distinctive aspect of migrant incorporation in Turkey (particularly in Istanbul), experiences of migrant illegality in Morocco have given rise to a search for political legitimacy.

Despite abuses and discrimination, irregular migrants have experienced legitimacy, in particular through their labour market participation in the case of Turkey. Migrant illegality has been absorbed in the informal urban economy. Rather lax control regimes in the urban context of Turkey have reinforced the informal incorporation process. However, this form of incorporation does not necessarily entail the recognition of irregular migrants' fundamental rights. The access to fundamental rights requires the possession of an official status. De facto recognition of the presence of irregular migrants is not coupled with an inclusive pro-migrant rights movement. Consequently, the rights of irregular migrants have not become a legitimate subject in political discussions.

In the case of Morocco, violence and discrimination characterize migrant illegality in daily interactions with security forces and within socio-economic life. Irregular migrants have had difficulties in finding employment opportunities and suffer from rights abuses by the police and street violence by locals. Their fundamental rights have been denied despite efforts by civil society and recent legislative changes. In the absence of day-to-day legitimacy, for instance through migrant participation in the labour market, migrant solidarity organizations have tried to mitigate exclusion in daily life. As civil society embraced the issue of rights violations of irregular migrants within Morocco, migrant organizations in collaboration with Moroccan and international civil society have gained a political voice. The result is increasing discussion of the vulnerable situation of irregular migrants in Morocco and increasing visibility of migrants and their associations claiming rights and legal status. For migrants, it has been a process of gaining visibility and legitimacy in the political sphere. Migrant associations, increasing the horizon of alliances with national, transnational civil society, and semi-public institutions, have publicly demanded regularization. Regardless of the outcomes of the regularization campaign initiated by the government in 2013, irregular migrants raising their voices in the public sphere and engaging in policy discussions with policymakers provide enough evidence to highlight migrant illegality in Morocco as a case of political legitimacy in the absence of economic incorporation.

The comparison highlights the contrast between the production of day-to-day legitimacy for migrants without a political voice in Turkey and the process of gaining political legitimacy without daily forms of inclusion in Morocco. The case of Turkey shows that the mechanisms through which irregular migrants gain legitimacy do not necessarily entail migrants' endeavours for political recognition. In other words, irregular migrants may not necessarily need to be political subjects to legitimize their presence. Meanwhile, daily forms of inclusion without political recognition reinforce their 'rightless' condition.

At this point, one can question the role played by market forces vs. pro-migrant actors in civil society in providing legitimacy for the presence of irregular migrants in society. However, it is too simplistic to conclude that there is a trade-off between these different forms of legitimacy. Instead, Chapter 6 questions, at a theoretical level, what kind of membership is envisaged by different mechanisms, providing legitimacy to the presence of irregular migrants in society in general and in new immigration contexts in particular.

6   Conclusions

6.1   Researching migrant illegality beyond externalization

There has been extensive media coverage of shipwrecks in the Mediterranean resulting in mass killings of migrants on their way to the EU.[1] Migrants have been filmed 'attacking' the fences surrounding Ceuta and Melilla.[2] There are reports of others switching tactics and using land, rather than sea borders to cross from Turkey into the EU, as Frontex operations are becoming stricter in the Aegean Sea.[3] The endless endeavour to stop irregular entry into the EU has resulted in the outsourcing of EU border security regimes into regions peripheral to the EU, especially in the Mediterranean basin. Increasing research has focused on technical investments to stop irregular border crossings in the Mediterranean and on the role played by smuggler networks (FRA 2013). The high death toll among migrants trying to cross the Mediterranean into the EU in the last decades (Brian and Laczko 2014), is a direct result of the fight against irregular migration by the EU.[4] Tens of thousands migrants fleeing conflict situations in Syria, Iraq, Afghanistan, and other countries were allowed to travel along the Balkan route to reach destinations in Western Europe. Reportedly, nearly a million people crossed during a short period in 2015 before the borders were closed down and securitized again.

Against the backdrop of these dramatic international events, which have turned international migration into a spectacle, this book has focused on

---

1   'Lampedusa boat disaster: Death toll rises to 232', *BBC News,* 07.10.2013, Retrieved 15.05.2015, from http://www.bbc.com/news/world-europe-24436779
'Migrant boat was "deliberately sunk" in Mediterranean sea, killing 500', *Guardian*, 15.09.2014, Retrieved 15.05.2015, from http://www.theguardian.com/world/2014/sep/15/migrant-boat-capsizes-egypt-malta-traffickers
'UN says 800 migrants dead in boat disaster as Italy launches rescue of two more vessels', *Guardian*, 20.04.2015, Retrieved 15.05.2015, from http://www.theguardian.com/world/2015/apr/20/italy-pm-matteo-renzi-migrant-shipwreck-crisis-srebrenica-massacre
2   'Over 1,000 migrants lined up along border ready to jump into Melilla', *El Pais*, 12.10.2012, Retrieved 15.05.2015, from http://elpais.com/elpais/2012/10/17/inenglish/1350489064_368373.html
3   'Greece Unnerved by Bulgaria's Schengen Prospect', *novinite.com*, 03.05.2011, Retrieved 15.05.2015, from http://www.novinite.com/articles/127902/Greece+Unnerved+by+Bulgaria%27s+Schengen+Prospect
4   'Missing Migrants Project' by IOM, http://missingmigrants.iom.int/. See also, 'The Deaths at the Borders Database' for a collection of official death toll at the EU borders, http://www.borderdeaths.org/

less visible outcomes of these developments unfolding at the edges of the EU and beyond its external borders since the 1990s. In recent decades, drastic changes have occurred in the governance of irregular migration in the Mediterranean basin at the borders of the EU and beyond. Consequences of restrictions on legal migration and the expansion of border controls beyond the EU are not limited to the closing of 'front doors' for (potential) migrants from the developing world, aspiring to better opportunities in life. This study has revealed the repercussions of these policy changes for migrant livelihoods at the periphery of Europe.

The initial motivation of the study was to explore the impact of the closure along the EU borders on migrant experiences of illegality at the periphery of these borders. The research questions have gone further to explore how migrant illegality is produced, practiced, and negotiated by state and non-state actors, migrants included. What is the impact of the new and external character of the production of migrant illegality on migrants' experiences? What are the implications of emerging forms of governance of irregular migration for migrant illegality, for migrant incorporation, and access to legal status? Under what conditions have those migrants who stay without legal authorization sought and gained social and political legitimacy? At a more theoretical level, what does the comparison reveal about the interconnection between the governance of irregular migration and the recognition of irregular migrants as rights-bearing subjects?

Taking migrant illegality as a constructed, thus reversible social condition, the analysis has unpacked the relationship between production of migrant illegality, migrant incorporation, and migrant mobilization. To explore different aspects of migrant illegality as a 'juridical status', 'socio-political condition', and 'mode of being-in-the-world' (Willen 2007a), I have engaged in three sets of research: on the legal production of migrant illegality, taking a socio-legal perspective on the question of illegality; on migrant incorporation, drawing on the sociology of migration in general with a specific focus on newcomers, especially those without legal status; and on migrant mobilization informed by the conceptual framework of social movements and contentious politics literature.

I have borrowed from the legal production of migrant illegality literature the idea that the category of 'illegal migrant' has been created by the law itself (De Genova 2002; Calavita 2005). Moreover, that this category of illegal has been sustained not only through immigration policy, but also through certain techniques of governmentality whereby migrant illegality is associated with criminality, racialized, and represented as a threat to national security. The production of migrant illegality making migrant bodies deportable, referring to the possibility of deportation rather than the actual practice of it, has given rise to mechanisms disciplining the migrant body. In a similar vein, research on migrant incorporation in the case of irregular migrants has revealed processes of subordinate inclusion into the society (Menjívar 2006; Bommes and Sciortino 2011). Conceptually and methodologically, I have drawn attention to layers of exclusion denying migrants' access to socio-economic opportunities and rights, as well as to socio-political and institutional conditions enabling migrants' participation in society, access to rights through the labour market, bureaucracy and/or civil society. Migrant incorporation research has also addressed how migrants themselves negotiate these conditions of exclusion and subordinate inclusion through tactics of (in)visibility. How irregular migrants become political subjects, actively seeking rights and claiming legal status, is also theoretically and empirically addressed by migration and social movements scholars alike (Coutin 2003; Nicholls 2013). The book has borrowed from research on social movements and its emphasis on the internal organization of the movement, repertoires of resistance, and alliances among diverse groups (Chimienti 2011; Tyler and Marciniak 2013). Acknowledging these factors underscoring the importance of institutional context, I put equal emphasis on migrants' lived experiences of illegality to explain processes leading to migrant mobilization and politicized visibility, as well as their lack of mobilization and subtle forms of invisibility.

Using these three interrelated research agendas, this book argued that what I have conceptualized as 'irregular migrants' incorporation styles' should be studied at multiple levels: at point where the legal framework, deportability practices, labour market and social conditions, the institutional context of bureaucracy and civil society receiving migrants meet. At the theoretical level, the research questions aimed to explore the relationship between (il)legality, recognition, and legitimacy. In exploring how migrants negotiate between state controls imposed upon them and their quest for formal recognition, the findings revealed the different ways that migrants gain legitimacy and political subjectivity despite their lack of legal status. Like other researchers studying lived experiences of illegality, I have considered a transformative social justice agenda without taking the 'nation-state and its interests at face value and as a point of departure' (Però and Solomos 2010: 11).

The research design embedded a number of empirical and conceptual novelties. The literature on irregular migrants' access to rights and legal status rarely focuses on contexts outside of North America and Western

Europe (Sadiq 2008; Garcés-Mascareñas 2012). Going beyond traditional geographies of comparison in migration research, the research questions explored external and domestic dynamics in the production of migrant illegality. Going beyond state-centric approaches to irregular migration, the book revealed the impact of this illegality on migrant experiences of subordinate incorporation and access to rights and legal status. Going beyond the focus on EU borders, the research provided a more comprehensive perspective than earlier research on the impact of EU migration control regimes on migrant rights at the periphery. Using the explanatory power of a comparative study, I aimed to transpose the emerging discussions on migrant illegality, incorporation, and legitimate access to the right to stay onto new immigration contexts.

The findings are based on data collected during fieldwork conducted at two research sites in the Mediterranean basin, Turkey and Morocco. The comparative research design was confined to the analysis of the emergence of different policies and practices in the governance of irregular migration between 2000 and 2014 in these two nation-state contexts. The analysis focuses on the post-2000 period because this was the time when the two countries had started to introduce new laws to manage asylum and immigration flows. This is the period when both countries moved from having no policy to rudimentary forms of governance in the realm of immigration, immigrants became more visible, and civil society organizations developed an interest in improving the situation of migrants and asylum seekers. These two nation-states have experienced similar transitions in their migration patterns from sending labour migration to Europe into de facto lands of destinations, whose borders are most subjected to the external dimensions of EU migration policies.

The book contributes to emerging literatures on irregular migration in the context of Turkey and Morocco by introducing a migrant illegality approach into these studies. The empirical discussions aim to contribute to academic research in Turkey and in Morocco as well as policy discussions, by bridging policy-oriented macro-level research on changing legal and institutional structures governing irregular migration with micro-level sociological studies on migrant communities, i.e. their lived experiences. Both country cases provide reflections on how migrant illegality has been produced as a result of the interaction of EU priorities to curtail irregular migration and changing national legal frameworks and practices at multiple levels. Despite their formal exclusion from the sphere of citizenship and the rights associated with it, migrants have carved out social and political spaces, albeit through different processes of incorporation. This study has provided a glimpse into how people excluded from the body of citizenship, actively or implicitly, claim their legitimate right to stay in the nation-state territory, despite their liminal legal status.

Despite limitations inherent in comparison, which sacrifice the depth of each case, the comparative research design has been fruitful. Empirical descriptions, in the sense of systematic process analyses of the cases, have elicited mid-range, context-bound causalities. Thick empirical description of cases enabled us to build causal narratives to explain the interlinked relationship between the production of migrant illegality, migrant incorporation, and access to rights in each country case and across cases. Hypotheses emerging from the comparison of these two cases require testing with a larger number of cases; this constitutes a limitation for the findings of the book.

The conceptual framework – postulating that migrant incorporation styles manifest themselves at the point of interaction between different aspects of migrant illegality – underpin my methodological choice to explore migrant illegality at multiple levels. Therefore, one methodological contribution of the research is that it embraces 'studying through approach' (Shore and Wright 2003: 14; Van der Leun 2003; Tsianos and Karakayali 2010) to link legal and institutional policy analysis on irregular migration and sociological- ethnographic methods on migrant livelihoods. Another methodological and empirical novelty of the research is its exploration of the causal linkage between migrant illegality, migrant incorporation, and mobilization in a comparative perspective, in contexts underexplored by previous research with a similar conceptual framework (Calavita 2005; Willen 2007a; Chimienti 2011; Lauthenbal 2007).

Both the studying-through approach and the comparative research design have enabled me to address a set of empirical questions emerging from migrant illegality and migrant incorporation literatures in a contextualized manner. Chapter 2 provided a comparative outlook on the impact of the externalization of the EU borders on Morocco and on Turkey. Chapters 3 and 4 linked processes that produce migrant illegality to processes of migrant incorporation into society and to processes conducive to their political mobilization. The comparative analysis of the two country cases in Chapter 5 provided preliminary explanations for why certain incorporation patterns have prevailed in one context and not the other. The concluding chapter reflects on the themes discussed in the book and the different patterns of migrant incorporation styles, taking into account international, national and local factors. It also reflects on the implications of recent developments for Europe and for peripheral countries in the post-2015 period.

6.2 Production of migrant illegality at the international and national levels

Since the 1990s, both Turkey and Morocco have become destinations for migrants from their wider regions, who come to these countries to work, seek asylum, study or with the initial intention of crossing into Europe. Despite differences in the volume of incoming flows and migrant profiles, in terms of source countries and migrant motivations, the similar and rather external emergence of irregular migration has made these migrant illegality contexts comparable. This process, which I have identified as the 'international production of migrant illegality' refers to techniques of governance that operate across national borders. In both contexts, these techniques have been mainly EU-initiated efforts, such as increasing joint investments in infrastructure along the external borders of the EU shared with third countries to stop irregular entries in the Union, increasing the visibility and activities of international and intergovernmental organizations, and changing the legal framework governing irregular migration.

The label 'transit country' has been used by the EU and by these countries to identify themselves; at this time, when the term transit does not have factual validity, it is becoming increasingly difficult for people to 'transit' through Turkey or Morocco to the shores of the EU without proper documents. At the same time, the term 'transit' has been internalized and instrumentalized by policymakers in Turkey and Morocco to deny the fact that they are becoming immigration countries and to avoid their obligations vis-a-vis foreign nationals on their soil. The first two aspects of the international production of migrant illegality – that is, developments concerning border infrastructure and the role played by international and intergovernmental organizations – have been similar in the two contexts. However, the link between domestic structures and EU demands, the direction of changes in the legal framework, the motivations behind these changes and the overall politicization of the issue, have differed from one context to another.

In both countries, the institutionalization of migrant illegality has been infused with international and domestic demands. Turkey has been pressured by the EU to cooperate in stopping irregular migration into the EU. At the same time, lax visa policies have been criticized for leaving a back door open for the legal entry, overstay, and further mobility of irregular migrants. Recent legal changes in Turkey indicate a shift in the migrant illegality regime that is 'nicer on asylum seekers and tougher on irregular migrants' (Tolay 2012: 53). Morocco has not become a hub for irregular labour migration to the same extent that Turkey has in the last two decades. Hence, border-related dimensions of migration governance have initially been more instrumental and dominant in the production of migrant illegality in Morocco.

Different levels of politicization of irregular migration by the state also indicate different technicalities of governance in the two contexts. Morocco has displayed higher levels of criminalization and politicization of irregular migration. In Turkey, the state has put less effort into creating (negative) public opinion on the later acknowledged fact that Turkey has become a hub for labour migration and asylum, as well for those attempting the journey to Europe. Keeping things low profile, legal and institutional changes are contextualized within the technicalities of the EU accession process by policies and researchers alike.

One conceptual contribution of the study is to socio-legal studies on migrant illegality, by bringing an international perspective into the analysis, but also indirectly to studies on the external dimensions of EU migration policies by providing a fresh lens on their analysis, with the extended use of the concept of the production of migrant illegality. The starting argument of the book has been that the production of migrant illegality in the periphery of the EU, defined as transit migration, has been a product of the interaction between the international context and domestic politics. The comparison indicates that migrant illegality as a judicial status has been a consequence of interaction between international and domestic contexts.

One major implication of the EU pressure to stop, control, and manage irregular migration for the contexts in question is the unprecedented occupation in categorizing incoming flows. One should note that Turkey and Morocco, as well as other countries in the region facing similar transit conditions, are asked to stop people from transiting to their next destinations, but they are also criticized for undermining the fundamental rights of migrants in their territory. Making transit spaces safe third countries, by pushing them to introduce functioning international protection systems, has been an EU priority. The downside of the increasing will to categorize incoming flows is a discrepancy between asylum seekers, as a group with legitimate access to rights and protection, and irregular migrants; although, in practice, both groups are enmeshed with one another and migrants can easily move from one category to the other. Hence, this research indirectly problematizes the implications of the concept of mixed migration by showing the diversity of experiences and the need for protection for all. This is an empirical contribution of the book which also has policy relevance.

Meanwhile, the EU is not the only dynamic in shaping immigration policies in both contexts. Such an approach reifies the idea of a transit country and characterizes countries at the EU's periphery as 'passive victims' of their geographies. Rather, examining the political, social, and institutional conditions within the receiving contexts, and migrant experiences of it, better explains the production and implications of migrant illegality as an interaction between domestic socio-political factors and foreign policy. Herein lie the empirical and methodological contributions of this study.

6.3   Migrant incorporation styles: The problematic role of the market

This study has acknowledged the EU's role as a supra-national actor that has an impact on the production of migrant illegality in its periphery. However, the emphasis is not on policies and practices produced by the EU, but on the policies and practices of the state, as negotiated by non-state actors, including migrants themselves, in the peripheral countries. These countries face similar immigration pressures, but I argue that differences in state and civil society responses to irregular migration, in terms of local configurations of migrant illegality, make a difference in migrants' experiences of incorporation. Different levels of politicization of irregular migration, hence differences in the conceptualization of migrant illegality, impact migrant incorporation styles. The presence of migrants, especially those without legal status call into question the original fiction of modern society by problematizing redistribution and recognition relations between state and citizen:

> Those who enter the administrative apparatus accept a certain degree of control over their actions as a method of obtaining the benefit of a certified identity [...]. The sociological interest in irregular migration is motivated by the chance to explore the other side of this exchange, as in a natural experiment, where the avoidance of controls is pursued through the renunciation of political recognition and legal protection. (Bommes and Sciortino 2011: 221)

Following this promise to study the experience of irregular migration, Chapters 3 and 4 have showed the processes that lead to different styles of migrant incorporation. In Morocco, the criminalization of irregular migration at the policy level, and at the level of public opinion, go hand in hand with migrants' daily experiences of deportability and other forms of exclusion, both in the informal settings along the border and in the urban space. The puzzling question of whether and under what conditions migrants may still seek legitimacy in the absence of labour market incorporation is addressed in the example of Morocco. Chapter 3 has already characterized the experiences of migrant illegality as a socio-political condition in Morocco, with limited access to labour market opportunities, a lack of a sense of legitimate presence due to daily experiences of deportation, and limited access to rights and then only with civil society acting as an intermediary. How these conditions influence migrants' modes of being-in-the-world as political subjects is another puzzle the book has attempted to solve.

Conversely, Turkey demonstrates migrants' experiences of daily inclusion without access to a political voice, especially for groups in urban centres such as Istanbul. In other words, lower levels of politicization of the presence of irregular migrants at the policy level resonate with lower levels of enforcement of internal controls on irregular migrants in the urban space. Interestingly, this situation of 'arbitrary tolerance', resulting from a gap between law and implementation, underscore migrants' lack of recognition, either as villains, or as victims. As a result, irregular migrants are not considered political subjects by policymakers, or by their potential allies in civil society. This double lack of recognition has left migrants to the mercy of the moral economy of the market, where there is widespread implicit consent for certain types of exploitation. Giving the deportation practices and precarious but inclusive labour market situation in Istanbul, lower levels of advocacy for rights of irregular migrants, migrant illegality, in the sense of ways of being-in-the-world, correspond to migrants' invisibility in the political sphere despite their relatively widespread presence in the socio-economic sphere.

Chapter 4 revealed that the labour market, coupled with somewhat tolerant deportation regimes, does provide a source of legitimacy for the presence of irregular migrants in Istanbul. Civil society underscored the arbitrary and, at times, unlawful practices of detention and deportation by the Turkish police. Meanwhile, migrants interviewed expressed less concern about being deported, especially when compared to their counterparts in Morocco, as long as they resist engaging in conflict at the workplace or in their neighbourhood. However, as the case of Turkey reveals, migrants' presence in the labour market does not necessarily provide a basis for their formal recognition. In other words, migrants gaining daily legitimacy through their economic participation and through practices of arbitrary toleration have not necessarily gained a political voice, hence formal recognition. Migrant illegality as a socio-political condition in Turkey has been characterized by subordinate forms of incorporation in the labour market, day-to-day legitimacy, and very limited access to fundamental rights without official status.

As immigration policies become part of the public policy agenda, and given the rapidly changing legal, institutional, and discursive contexts on irregular migration in both countries, it would be interesting to explore how these policies interact with other public policies on health, education, and the labour market. Related research may put more emphasis on the functioning of bureaucratic incorporation and focus on the perspectives of 'street level bureaucrats' (Lipsky 1980), such as the police, doctors, and school principals. Generating institutional ethnographic analysis that reveals mechanisms for accessing rights and the role played by street level bureaucrats would help us to theorize on the hierarchy of values in vulnerable groups' (including those without formal membership) access to the institutions of the nation-state (Fassin 2009). In addition, the role of non-state actors, further analysed and problematized from an empirical and theoretical perspective, would be beneficial as would looking at the role of civil society in perpetuating or mitigating the functioning of emerging hierarchies. The latter would be in dialogue with existing literature on problematizing CSOs taking responsibilities that are conventionally undertaken by states and what this means for recognition and redistribution of resources.

6.4   Migrant mobilization between (in)visibility and recognition

The emergence of migrant political mobilization is an empirical question rather than an intrinsic aspect of migrant illegality, both as a socio-economic condition, but also as a mode of being-in-the-world. My research began with the premise that migrants are not only victims of external conditions, but active subjects even in inhospitable contexts characterized by violent practices. They consciously endeavour to improve their living conditions within the political, social, and institutional constraints and opportunities surrounding them. One central question of the thesis has been the difference at the level of mobilization for the rights of irregular migrants in the context of Turkey and Morocco, despite similar experiences of being stranded, the denial of rights, and the experience of violence in both contexts, as reported by several international and local civil society actors.

Exclusion has created a situation where migrant rights have been denied, irregular migration has been criminalized, and irregular migrants have been stigmatized in Morocco, arguably at higher levels than Turkey. As a response to stigmatization at different levels, irregular migrants' mobilization has become part of their way-of-being in the world. Their common African identity and linguistic background has facilitated migrants' communal quest for political recognition. Forging alliances with emerging pro-migrant rights civil society actors in the Moroccan context, irregular migrants have been partially successful in carving a political space where they contest their illegality characterized by a lack of judicial status and as socio-economic conditions excluding them.

Using a political opportunity structures approach, albeit not exclusively, I have connected migrants' mode of being-in-the-world as political subjects in Morocco to their socio-economic conditions, but also to wider institutional structures. Socio-political conditions of migrant illegality characterized by marginalization in the absence of day-to-day legitimacy have, arguably, put irregular migrants in Morocco into a much more vulnerable position than their counterparts in Turkey. The main factor enabling migrants' grievances to transform into contentious politics in Morocco has been informal migrant associations' alliances with Moroccan and international civil society. The conscious decision by Moroccan civil society actors not to distinguish between asylum seekers vs. irregular migrants, as, for instance, UNHCR Morocco would prefer, has arguably carved out a wider political space for contentious politics. In the case of Turkey, migrants and asylum seekers have mobilized among themselves in sporadic ways. Apparently, in the Turkish case, most NGOs prefer to focus on asylum as a more legitimate basis for their advocacy activities, at the expense of sidelining rights violations and the protection needs of irregular migrants.

In both contexts, discussions of irregular migration were initially shaped in relation to clandestine outmigration of their own nationals. Emigration and the situation of migrants from Morocco are still relevant for policy discussions in Morocco. As part of transnational opportunity structures (Però and Solomos 2010: 9-10), these discussions, as well as institutions dealing with emigrants abroad, have had an impact on ideas of how immigrants within Morocco should be treated (Üstübici 2015). This linkage between issues pertaining to emigration and immigration has so far been absent in Turkey, despite its similar emigration history. In this sense, the comparison reveals that the *success* of mobilization does not depend solely on the existence of allies supporting demands by marginalized, not formally recognized groups, but also how these allies formulate their priorities, where they think they can find discursive opportunities (Bröer and Duyvendak 2009). Overall, the causal assessment for migrant (im)mobilization throughout this book suggests that the mobilization of migrants without legal status can be explained by a combination of various factors, such as the perception of closure in the system (Chimienti 2011). Here, the closure in the system refers to the actual and ongoing closure of EU borders, which makes the possibility of exit more costly and risky. But it also refers to the intensity of internal controls rendering irregular migrants' efforts to lead a decent life much more difficult. The sense of closure coupled with migrants' own background of political mobilization, leading to the possibility of forming coalitions with other actors, helps to transform migrants' personal experiences of harm into wider political demands for recognition (Honneth 1995: 163).

One potential contribution to the literature at the intersection of irregular migration and social movements lies in the conviction that cases of mobilization are just as useful as those identified with a lack of mobilization for generating hypotheses. Given the similar international dynamics in the production of migrant illegality in both contexts, one can ask under what conditions migrant illegality translates into cheap, flexible labour. The case of Turkey has already revealed that the availability and penetrability of the urban informal economy, the size of the economy, as well as the existence of already emerging ethnic economies, which would welcome new (irregular) migrants, are factors enabling the creation of an inexpensive, vulnerable migrant labour force in Turkey. Surprisingly, access to precarious work has been the case even for those migrants who are allegedly on their way to Europe. More important for this research, the other puzzling question is whether or under what conditions irregular migrants' subordinate incorporation into the labour market may turn into a mechanism that weakens migrants' quest for formal recognition. Most literature in Western Europe emphasizes the economic contribution as the basis for political demands of legalization of irregular migrants. As hypothesized in Chapter 5, the labour market participation is neither necessary nor sufficient for the political mobilization of irregular migrants.

6.5   Ways forward

The conceptual framework I proposed regarding the international and internal production of migrant illegality may be extended, confined, and refined by three types of further research: by transposing it onto other countries emerging as de facto lands of immigration at the edge of conventional destinations, by following changes on migration governance and migrant incorporation patterns over time, by shifting the scope of the analysis within each case explored here. More research on the creation and transformation of so-called transit spaces into migrant receiving lands is needed to extend the engagement of this book with changes to legal and institutional frameworks and to further theorize the kind of incorporation styles that the transformation of migrant illegality regimes has given rise to.[5] Such an inquiry will enable us to capture the variation among similar cases. It will also highlight the dynamics of the journey and settlement from the perspective of migrants.

Taking into account the policy implications of the book, further research is needed to assess the extent to which changing legislation in Morocco and Turkey would provide necessary protection and rights and disincentivize migrants from furthering the journey as envisaged by the EU, or would result in further illegalization of certain types of mobility. Looking at the recent changes in both contexts, the findings of this book can be extended into an analysis of how the (international) production of migrant illegality has evolved into frames of 'deservingness'. Such an inquiry would contribute to answering the question of what makes a foreigner a deserving citizen (Chauvin and Garcés-Mascareñas 2014).

The findings of this research are confined to the emergence of migrant mobilization and initial phases of its formal recognition by the authorities in Morocco. At the end of the first regularization campaign in Morocco, only a limited number of irregular migrants had acquired an exceptional temporary residence permit and coercive practices along the EU border continued throughout 2014. After the end of the regularization in 2015, the Moroccan government engaged in policies to ensure the integration of migrants regularized in Morocco. Since 2014, the Western African route into the EU has been virtually closed, but migrants from sub-Saharan countries continue to arrive in Morocco. As it is not clear how migrants whose applications are rejected and those who arrived after the campaign will be treated, the Moroccan government has initiated a second regularization campaign. Thus, more comprehensive research is needed on the evolution of migrant mobilization in Morocco. What inclusionary and exclusionary patterns will arise within the movement, as well as with similar movements in the country and in the wider region is yet to be seen and researched.

Needless to say, the mass arrival of nearly 2.8 million Syrians as of November 2016, who were granted temporary protection status in Turkey, has significantly changed the scene of migrant incorporation and migration governance in Turkey. During this time, attention shifted from irregular borders crossings from the Western African route to the Eastern Mediterranean route. Turkey's reception of high numbers of Syrian refugees has reinforced migration diplomacy with the EU. The Turkey-EU joint statement of March 2016, the so-called Turkey-EU deal, should be analysed as a continuation

---

5   See for instance Basok and Rojas Wiesner 2017 for the impact of regularization programs in Mexico.

of the trend towards externalizing the EU's borders and migration policies and of migration diplomacy between Turkey and the EU.

The patterns of politicization of immigration issues, easy infiltration into the labour market, and daily forms of tolerance have started to change in Turkey, perhaps not unexpectedly in any society receiving nearly three million newcomers over a period of five years. It is fair to say that the interaction of the Syrian refugee situation with ongoing legal and administrative changes in migration governance in Turkey have gradually created hierarchical categories of deservingness and reinforced the illegality of those who fall outside these categories. With the widespread employment of Syrians in the labour market, Turkey is now becoming an even better case for analysing how economies react to the entry of a cheap migrant workforce without the explicit demand for labour. Here, further research can focus on the community characteristics of incorporation. The findings can also be extended into more systematic research about the intersectionalities of gender, class, ethnicity, and race and how they differentiate experiences of illegality. What kind of solidarities and networks emerge through these experiences?

On the one hand, the migrant illegality framework is not sufficient to address Syrian refugees, who are under Temporary Protection, hence not illegal. On the other hand, initial research has already revealed a discrepancy between the recognized legal status of Syrians and their lived experiences of incorporation, displaying parallels with living conditions of irregular migrants, especially with regards to labour market participation (see for instance Özden 2013; Korkmaz 2017). Thus, this study provides a basis for research on the incorporation of Syrians, not only to help to illustrate the drastic changes that have occurred, but also to reveal continuities in patterns of migrant incorporation in Turkey. The impact of the mass arrival of Syrian refugees, as well as the increasing number of asylum seekers from other countries, has put Turkey's asylum system under stress. Whether the increasing number, coupled with increasing visibility in political discussions, will translate into a form of political mobilization in Turkey is yet to be seen. One speculative question would be whether an asylum-based mobilization would expand to include other categories of migrants making more radical membership claims and demands for free circulation.

In Morocco and in Turkey, as well as elsewhere, more changes are needed to ensure that migrants, regardless of legal status, gain access to their fundamental human rights. As repeatedly uttered by the migrant activists I interviewed: *la lutte continue* ('the struggle goes on').

Annex

Table 1 Interviews with state institutions – Turkey

Name of Institution	Place of Interview	Date of Interview
Ministry of Interior- Bureau of Migration and Asylum	Ankara	February 2012
Ministry of Labor and Social Security- Bureau of Work Permits for Foreigners	Ankara	February 2012
Ministry of Foreign Affairs- Directorate General for Migration, Asylum and Visa	Ankara	November 2012
Ankara General Directorate of Security	Ankara	December 2012

Table 2 Interviews with international organizations, non-governmental organizations – Turkey

Name of the Institution	Place of Interview	Date of Interview
Association for Solidarity with Asylum Seekers and Migrants (ASAM)	Ankara	December 2012
International Organization for Migration	Ankara	December 2012
Amnesty International	Ankara	November 2012
Human Resource Development Foundation (HRDF)	Ankara	December 2012
Human Resource Development Foundation (HRDF)	Istanbul	August 2013
Helsinki Citizens Assembly (HCA)	Istanbul	November 2012
Caritas	Istanbul	December 2012
Doctors without Frontiers (MSF)	Istanbul	November 2012
Association for Human Rights and Solidarity with the Oppressed (Mazlumder)	Istanbul	November 2012
Migrants' Association for Social Cooperation and Culture (Göç-Der)	Istanbul	November 2012
Association for Solidarity and Mutual Aid with Migrations (ASEM)	Istanbul	November 2012
Foundation for Society and Legal Studies (TOHAV)	Istanbul	November 2012

Table 3   Interviews with migrants – Turkey

Nickname	Age	Sex	Nationality	Legal Status	Employment	Date of Interview
Rakel	50s	F	Armenia	Legal entry – Violation of visa	Cleaning lady	January 2012
Alima	34	F	Eritrea	Undocumented entry- asylum seeker – refugee	Interpreter	March 2012-November 2013 (multiple interviews)
Maria	20s	F	Ethiopia	Undocumented entry	Cleaning lady	March 2012
Nicole	20s	F	Democratic Congo	Undocumented entry – asylum seeker application	Unemployed	March 2012
Chris	36	M	Nigeria	Legal entry - Violation of visa – residence permit hrough amnesty	Unemployed, textile	June 2012- March 2013 (3 interviews)
Mahmut	29	M	Afghanistan	Undocumented entry	Textile-construction	January 2013- August 2013 (2 interviews)
Ahmet	22	M	Afghanistan	Undocumented entry	Textile-construction	January 2013- March 2013 (2 interviews)
Malik	22	M	Afghanistan	Undocumented entry	Textile-construction	January 2013-August 2013 (3 interviews)
Harun	22	M	Afghanistan	Undocumented entry –applied for residence permit	Textile	August 2013
Afsana	37	F	Afghanistan	Undocumented entry –consider applying asylum	Textile, spouse unemployed, then construction	February 2013- August 2013 (2 interviews)
Selma	32	F	Afghanistan	Undocumented entry –acquired residence permit	Textile, Spouse unemployed then, construction	February 2013- August 2013 (2 interviews)
Zerrin	Late 30s	F	Afghanistan	Undocumented entry -Applied for asylum- lives Istanbul- later moved to satellite city	Textile, work with children	March 2013
Shiba	30	F	Afghanistan	Applied for asylum	Textile, work with sibling	March 2013
Muzaffar	44	M	Pakistan	Undocumented -asylum applicant	Unemployed, wants to go to Europe	March 2013
Onur	28	M	Iran	Legal entry (?), Applied for asylum- Istanbul	Owns a restaurant	February 2013
Blessing	Late 30s	F	Nigeria	Legal entry –residence permit- Overstay	Unemployed, sex worker, textile	June 2013-August 2013 (2 interviews)
Dilbar	33	F	Uzbekistan	Legal entry, Violation of visa, passport has been stolen	Domestic and care work	March 2013
Katerin	43	F	Moldova	Legal entry, acquired residence permit, violation of visa	Domestic and care work	April 2013
Fatma	45	F	Uzbekistan	Legal entry, Violation of visa	Domestic work	April 2013

Nickname	Age	Sex	Nationality	Legal Status	Employment	Date of Interview
Rabia and her brother Halim	35	F	Afghanistan	Legal entry, Violation of visa, papers stolen, apprehended, asylum application, voluntary return	Unemployed	March 2013
Feriye	22	F	Afghanistan	Undocumented entry, residence permit	Textile	August 2013
Peter	34	M	Nigeria	Legal entry, Violation of visa, amnesty, Violation of visa	Textile	September 2013
Victor	29	M	Georgia	Legal entry-exit, residence permit, Violation of visa	Various petty jobs, hotel, restaurant	September 2013
Uche	27	M	Nigeria	Legal entry, Violation of visa	Unemployed, textile	September 2013
Alex	33	M	Nigeria	Legal entry, Violation of visa	Unemployed, textile	September 2013
Halim	34	M	Syria	Legal entry – acquired residence permit	Tourist guide	November 2011
Sultan		F	Azerbaijan	Legal entry –acquired residence permit-citizenship	Unemployed, spouse in textile	November 2011
Eric	34	M	Cameroon	Legal entry, Violation of visa	Textile	November 2013
Jackie	28	F	Ethiopia	Legal entry, residence permit, Violation of visa, asylum	Masseuse	November 2013

Table 4   Interviews with state institutions – Morocco

Institution	Place of Interview	Date of Interview
Hassan II Foundation	Rabat	April 2012- June 2012 (2 interviews)
Ministry of Education Directorate for Rabat-Sale Region	Rabat	July 2012
Deputy in Moroccan Parliament	Rabat	September 2012
The Council of the Moroccan Community living abroad (CCME)	Rabat	July 2012- September 2012 (2 interviews)
National Council for Human Rights (CNDH)	Oujda	September 2012
Ministry in Charge of Moroccans Abroad and Migration Affairs	Rabat	July 2012- May 2014 (2 interviews before and after the reform)

Table 5 Interviews with International Organizations, NGOs – Morocco

Institution	Place of Interview	Date of Interview
Moroccan Organization of Human Rights (OMDH)	Rabat	April 2012
Moroccan Association of Human Rights (AMDH)	Oujda – Rabat	September 2012 (2 interviews)
The Anti-racist Group for the Support and Defence of Foreigners and Migrants (GADEM)	Rabat	April 2012 – May 2014
Association Rencontre Méditerrannènne pour l'Immigration et le Développement (ARMID)	Tangiers	April 2012 – July 2012
Beni Znassen Association of Culture, Development and Solidarity (ABCDS)	Oujda	September 2012
East-West Foundation (FOO)	Rabat	April 2012 – September 2012 (3 interviews)
Doctors without Frontiers (MSF)	Rabat – Oujda	April 2012 – September 2012 (2 interviews)
Terre des Hommes (People of the Earth)	Rabat	April 2012 (2 interviews)
CARITAS	Rabat – Tangiers – Casablanca	July 2012 – September 2012 (3 interviews)
Center for Welcoming Migrants (SAM)	Casablanca	September 2012
The UN Refugee Agency (UNCHR)	Rabat	April 2012 – May 2014 (2 interviews)
International Organisation for Migration (IOM)	Rabat	May 2014
Democratic Organization of Labour – immigrant workers (ODT-IT)	Rabat	September 2012 – May 2014 (2 interviews)
Council of Sub-Saharan Migrants in Morocco (CSMM)	Rabat	September 2012
Collective of Sub-Saharan Migrants in Morocco	Rabat	September 2012 (2 interviews)
ALECMA (Association Lumiere sur l'emigration clandestine au Maghreb)	Rabat	May 2014

Table 6 Interviews with migrants – Morocco

Nickname	Age	Sex	Nationality	Legal Status	Employment	Date of Interview
Oumar	22	M	Guinea	Legal entry- Overstay	Unemployed- football player	July 2012 (Rabat)
Amadou	26	M	Senegal	Legal entry- student visa- started working	Proximity agent	July 2012- September 2012 (Rabat)
Modou	33	M	Senegal	Legal entry- started working- work permit	Masseur	July 2012 (Rabat)
David	29	M	Guinea	Undocumented entry	Unemployed- football player	July 2012 (Rabat)
Moussa	56	M	Guinea	Entry with passport-overstay – illegal border crossing- residence permit via marriage	Unemployed	July 2012 (Rabat)
Hafiz	40?	M	Cameroon	Entry with passport- overstay	Unemployed	July 2012 (Rabat)
Khadim	24	M	Senegal	Student-started working	Call center	July 2012 (Rabat)
Elou	28	F	Senegal	Legal entry- work permit expired	Stay-in domestic worker	July 2012 (Rabat)
Amy	27?	F	Philippines	Legal entry- residence permit- overstay	Domestic worker	July 2012 (Rabat)

Nickname	Age	Sex	Nationality	Legal Status	Employment	Date of Interview
Alassane	35	F	Cameroon	Undocumented entry- trying to go to Europe	Unemployed	July 2012 (Tangiers)
Adama	18	M	Senegal	Legal entry-exit, trying to go to Europe	Unemployed	July 2012 (Tangiers)
Cherif	21	M	Senegal	Legal entry-exit, trying to go to Europe	Unemployed	July 2012 (Tangiers)
Demba	22	M	Senegal	Legal entry-exit, residence permit in Spain, sentenced in Morocco for forgery	Unemployed	July 2012 (Tangiers)
Issa ve Yaya	32-24	M	Guinea	Undocumented entry- trying to go to Europe	Unemployed	July 2012 (Tangiers)
Yassine	24?	F	Senegal	Legal entry-exit- overstay	looking for a job, braiding hair	September 2012 (Casablanca)
Jules	37	M	Republic of Congo	Undocumented entry	Unemployed	September 2012 (Rabat)
Angela	42	F	Philippines	Legal entry- overstay	Cleaning lady, masseuse, hairdresser	September 2012 (Rabat)
Danny	30s	M	Nigeria	Undocumented entry	Construction, unemployed	September 2012(Rabat)
Anna	27	F	Democratic Republic of Congo	Undocumented entry- on a wheelchair, application for asylum	Unemployed, disabled	September 2012 (Oujda)
Rosa	42	F	Democratic Republic of Congo	Undocumented entry- recognized refugee	Runs a woman coop-erative, handcraft	September 2012 (Rabat)
Maria and her elder sister Edith	29-45	F	Democratic Republic of Congo	Undocumented entry	handcraft, hair braiding, sex worker	September 2012 (Rabat)
Amar	23	M	Niger	Student visa, overstay	Unemployed, was looking for a job, going to his country	September 2012 (Rabat)
Maya	22	F	Guinea	Short term student visa	Unemployed, ODT volunteer	September 2012 (Rabat)
Mama	52	F	Ivory Coast	Asylum application	Unemployed, handcraft	September 2012 (Rabat)
Linda	42	F	Chad	Student-residence permit	Works in associations, wants to start one of her own for women	September 2012 (Rabat)
Oumar	32	F	Guinea	Legal entry, residence permit with short term student visa	Works in associations	September 2012 (Rabat)
Patrik	33	M	Cameroon	Legal entry, residence permit with short term student visa, wants to go to Europe	Petty jobs, works in associations	September 2012 (Rabat)
Fatima	20s	F	Nigeria	Undocumented entry, saves money for going to Europe	Begging	September 2012 (Rabat)
Papa	28	M	Ivory Coast	Undocumented entry	Unemployed	September 2012
Jean Baptiste	25	M	Democratic Republic of Congo	Undocumented entry	Unemployed	September 2012 (Rabat)
Sunny	38	M	Nigeria	Undocumented entry, Asylum application	Unemployed, petty jobs, begging	September 2012 (Rabat)
Naima	29	F	Central African Republic	Undocumented entry, Asylum application	Unemployed	May 2014 (Rabat)
André	42	M	Cameroon	Undocumented entry, Asylum applica-tion, applied for regularisation	Petty jobs in construction, carrier, voluntary work in associations	May 2014 (Rabat)

Table 7   Immigration flow into Turkey and Morocco

Legal category		Morocco	Turkey
irregular migration	Source countries	Senegal Nigeria Ivory Coast Guinea Congo Mali The Philippines Cameroon DRC Syria (based on previous surveys (AMERM, 2008) and registrations for the **regularisation campaign in 2014.**	Afghanistan Burma Eritrea Pakistan Iraq Georgia Turkmenistan Azerbaijan (based on apprehended cases in **2014** as reported by DGMM)
	Estimated number	The number of apprehended cases fluctuates between 10,000 and 20,000 per year. It was as high as **23,851** in 2003 (Khachani, 2011: 4). During **2014, 27,330** migrants without legal status in Morocco registered for the **regularisation** campaign.	The number of apprehended cases fluctuates between **29,926** in **1998** and **58,647** in **2014.** It was as high as **94,514** in **2000** (as reported by DGMM).

Legal category		Morocco	Turkey
Asylum	Source countries	Congo-Kinshasa Ivory Coast Syrians Mali Nigeria Cameroon	Iraq Afghanistan Iran Somalia
	Estimated number as reported by *UNHCR*	**3580** (total population of concern by **April 2015**)	**2935** (in **2005**) **56,709** (registered active case-load by January 2015 excluding more than 1.5 million **Syrians** under temporary protection by the end of **2014**.)
Legal Residents	Source countries	France Algeria Spain Senegal Mauritania US	Iraq Syria Afghanistan Iran Russian Federation Turkmenistan Germany UK Georgia (based on residence permits issued in 2014)
	Estimated number	**77,798** (2012) (as reported by Moroccan Directorate General of National Security).	Increased from **182,301** in 2010 to **379,804** in 2014 (as reported by DGMM).

Table 8   Migration policies in Morocco and Turkey (2000-2014)[1]

Morocco

Date	Policy	Notes on the content / context / intended impact
1999	*The 1999 Action Plan proposed by the High Level Working Group on Asylum and Migration*[2]	The Action Plan identified Morocco as a "transit country" and envisaged legal, infrastructural changes. The Plan was rejected by the Moroccan government.
2000	*Association Agreement with the EU*	Signed in 1995, entered into force in 2000. Both parties have agreed to cooperate on illegal migration.
2002	*Establishment of SIVE (Sistema Integrado de Vigilancia Exterior) along the Morocco–Spain border*	
2003	Law n° 65-99 relative to the Labour Code	Regulations on the employment of foreign workers, requiring authorisation by responsible government agency (Art. 516) and sanctioning the employment of foreigners without authorisation.
2003	*Readmission agreement with Spain*	The readmission agreement with Spain was ratified only in 2012. The implementation has been problematic and excluded third country nationals. Throughout 1990s, Morocco signed readmission agreements with France (1993, 2001), Germany (1998), Italy (1998, 1999), Portugal (1999). Those agreements concern the return of Moroccan nationals.
2003	Law n° 02–03 relative to the entry and stay of foreigners in Morocco and to irregular emigration and immigration	– The law envisaged the institutionalization of a Directorship for Migration and Surveillance of Borders within the Ministry of Interior. – More investment in border surveillance – Criminalization of irregular migration of nationals and foreigners and its assistance – Article 26 provides legal basis and procedures to follow for detention and removal to the border. The article contains protective measures prohibiting the deportation of asylum seekers, refugees, pregnant women, and minors. – Regulating residence permits

1   The preparation of Annex 2 drew on data from the DEMIG Migration Policy Database collected within the DEMIG Project and funded by the European Research Council under the European Community's Seventh Framework Programme (FP7/2007-2013) / ERC Grant Agreement 240940 (see DEMIG, 2015).

2   Policy changes directly resulting from relations with the EU or concerning external borders of the EU are indicated *in italic*.

Date	Policy	Notes on the content / context / intended impact
2005	Decree n°1391-05 on employment of foreigners	ANAPEC (National Employment Agency) checks whether there is an eligible candidate with Moroccan citizenship amongst applicants (refugees are, on paper, exempted from labour market test).
2005	*Stricter control along the Ceuta and Melilla borders*	In response to Ceuta and Melilla events in October 2005, where official numbers indicate the killing of 11 migrants by Moroccan and Spanish border guards.
2005	Repatriation of around 3,000 irregular migrants back to their countries of origin	As a result of strengthened internal controls in the aftermath of Ceuta and Melilla events.
2005	Note no : 93 on the enrolment of foreign children in public education institutions	Provincial delegations of the Ministry are defined as the authority to decide on the school enrolment of children from other nationalities.
2006	First Euro-African Ministerial Conference on Migration and Development organized in Rabat	The aim of the conference was to establish a global dialogue on migration. Morocco as the organizer assumed a role of mediator between Northern and Southern countries.
2006	*Country Agreement with IOM*	
2007	*The Headquarter Agreement with UN High Commission for Refugees (UNHCR)*	Refugee status determination is processed by UNHCR. However, until the end of 2013, recognized refugees did generally not have automatic access to residence permits and other benefits such as health care, education.
2007	Law n°62-06 modified Dahir n°1-58-250 on the Citizenship Code.	The change has enabled the transmission of Moroccan citizenship for children born to Moroccan mothers and their foreign spouse.
2007	The Council for Moroccans Abroad (CCME) was established.	The institution aim at strengthening the dialogue between Moroccan state and Moroccan community residing abroad. Article 163 of 2011 Constitution recognized the role of the Council. The Council is constituted of representatives of the Moroccan community abroad, most of them are appointed by the King himself. The Council recently revealed an interest to immigration issues.
2010	Decree n°2-09-607 on the implementation of Law no:02/03	The implementing directive specifies terms and procedures for the provision of resident permits.
2011	*Morocco accepted the UN Protocol to Prevent, Suppress and Punish Trafficking in Persons, Especially Women and Children.*	Related changes in national legislation are still under preparation.

Date	Policy	Notes on the content / context / intended impact
2011	New Constitution	– Article 30 guarantees rights to foreigners within the country including the right to vote in local elections. The Article articulates that procedures for asylum and refugee determination should be determined by law. – There were no references to the rights of foreigners in the previous Constitution enacted in 1995.
2011	Law 34-09 relating to the "Health System and Offer of Care"	Based on the Circular in 2003, the Moroccan legislation recognises irregular migrants' right to health care.
2013	*The Mobility Partnership between Morocco, the EU and six EU Member states*	Further cooperation on co-development, combating irregular migration, administrative support for enacting asylum legislation in Morocco is envisaged.
2013	Report of the National Council of Human Rights (CNDH) "Foreigners and human rights : for a radically new policy"	CNDH invited government to take necessary legal measures to ensure human rights of migrants in Morocco.
2013	A new mandate given to the Moroccan Ministry for Moroccans living abroad	The name was changed into the Moroccan Ministry for Moroccans living abroad and Migration Affairs. A new department was founded to coordinate regularisation campaign and to work on new legislations on asylum, integration, human trafficking.
2013	Reopening of the bureau of refugees	The Moroccan state has started to process asylum files in collaboration with UNHCR.
2013	Circular n°13-487, 9 October 2013, concerning the access to education of migrant children from the sub-Saharan and Sahel regions	Accordingly, migrants can enrol their children to private and public schools in the country, regardless of legal status. Bureaucratic procedures for school enrolment have been simplified.
2013	The introduction of a health insurance scheme of social assistance (RAMED)	No provisions had been envisaged on the inclusion of foreigners in the system.
2014	An exceptional regularisation campaign for irregular migrants from Europe and Africa.	The campaign lasted throughout 2014. Eligibility: The exceptional operation of regularization concerns foreigners with spouses from Moroccan nationality living together for at least two years, foreigners with foreign spouses in legal status in Morocco and living together for at least four years, children from the two previous cases, foreigners with employment contracts effective for at least two years, foreigners justifying five years of continuous residence in Morocco, and foreigners with serious illnesses who had arrived the country before December 31 2013.

Date	Policy	Notes on the content / context / intended impact
		Outcome: "Close to 17,918 one-year residence permits were granted from 27,330 applications registered (almost half of them to Senegalese and Syrians, followed by Nigerians and Ivoirians)" (Martin, 2015)

Turkey

Date	Policy	Notes on the content / context / intended impact
1994	The Regulation on the Procedures and the Principles Related to Mass Influx and the Foreigners Arriving in Turkey either as Individuals or in Groups Wishing to Seek Asylum either from Turkey or Requesting Residence Permits with the Intention of Seeking Asylum from a Third Country, referred as 1994 Regulation.	– First national legislation on asylum and on how to implement 1951 UN Convention. – The Ministry of the Interior became the final decision-making body for refugee status determination in collaboration with the UNHCR. – The Regulation introduced administrative procedures requiring applicants to register with the police within five days of arrival and to reside in cities designated by the police.
2001	*Accession Partnership Agreement with the EU*	
2001	*Visa requirements for a number of states – including Kazakhstan, Bahrain, Qatar, the United Arab Emirates, Kuwait, Saudi Arabia and Oman in 2001 and 2002.*	In line with the EU visa policy
2002	*UN Conventions Against Transnational Organised Crime and its additional protocols*	Related changes in national legislation are introduced. The Law on Protection of Victims of Human Trafficking is still under preparation.
2002	*The Turkish National Security Council adopted a resolution on combating irregular migration in 2002*	
2003	*Readmission Agreement with Greece*	The implementation has been problematic and numbers have remained low. Other RAs were signed with Syria (2003), Kirghizstan (2004), Romania (2004), Ukraine (2005), Russian Federation (2010), Pakistan (2011), Bosnia Herzegowina (2012), Moldova (2012)

Date	Policy	Notes on the content / context / intended impact
2003	Law No. 4817 on Work Permits for Aliens	– Ministry of Labour was given the main responsibility for issuing work permits (Art. 3). – Temporary work permits are issued for a period of maximum one-year, by taking into account the conditions of the labour market and the availability of a better qualified Turkish nationals capable of performing the job (Article 5). – Article 21 introduced fines for both employers and migrant workers who are in an irregular employment situation.
2003	*Strategy Paper for the Protection of External Borders in Turkey*	Border management issues have been on the agenda concurrently with membership talks, along with migration management and asylum issues.
2003	Amendment in the Citizenship Law	– Article 5 was changed to introduce equal citizenship rights for foreign men and woman marrying Turkish citizens – Both men and women marrying Turkish nationals will have to wait three years to apply for Turkish citizenship. – The legislative change aims to prevent *marriages of convenience* by foreign migrant women, resulting in the acquisition of Turkish citizenship.
2004	*Turkey has become full member of IOM.*	– The activities of IOM in Turkey were initiated in 1991 in the aftermath of the regional crisis in the Middle East. – A bilateral agreement was signed in 1995. – Turkey became a full member of IOM in 2004 in the context of a national action plan on asylum and migration.
2004	*Asylum-migration twinning project with Denmark and the UK*	The outcome was the preparation of National Action Plan for Asylum and Migration launched in 2005
2005	*National Action Plan*	EU membership formally opened. – Three main issues need to be addressed by the Turkish government during the accession process were developing asylum legislation; signing readmission agreements with third countries; lifting the geographical limitation to the 1951 Refugee Convention. – Turkey committed to prepare the Law on Asylum, the Law on Aliens and to initiate the legislative process for lifting the geographical limitation by 2012.

Date	Policy	Notes on the content / context / intended impact
2005	Article 79 of the Turkish Penal Code (Law No:5237) was changed.	– Harsher penalties were introduced for human smuggling.
2006	Settlement Law No. 5190 is introduced replacing the 1934 settlement law which generally restricts immigration to persons of "Turkish descent and culture".	Although Settlement Law was amended in 2006 parallel to the EU harmonization efforts of Turkey, the new Law on Settlement still sustains this conservative nature and provides for individuals of "Turkish descent and culture" to be accepted as immigrants and refugees in Turkey.
2006	The 2006 Regulation on asylum clarifying the 1994 Asylum Regulation	The requirement of registration with official bodies within five days of arrival was removed.
2007	*Action Plan on Integrated Border Management*	– An outcome of the twinning project in collaboration with the UK and France. – A civilian body to protect the borders was envisaged.
2007	Opening of Kumkapı Removal Centre as a "Foreigners' Guesthouse"	Turkey committed to increase the capacity and numbers of removal and reception centres. EU partly funds the construction of these centres.
2008	Opening of Migration and Asylum Bureau under the Ministry of Interior	The main mandate was to work on drafting the law on asylum and foreigners.
2008	Opening of the Bureau for Border Management under the Ministry of Interior.	In the context of the Action Plan on Integrated Border Management
2009	Turkey's already liberal visa regime was further relaxed	– Agreements signed for visa-free mobility between Turkey and Syria (as of October 2009), Georgia (as of February 2006), Lebanon (as of January 2010), Jordan (as of December 2009) and Russia (as of May 2010). These changes have reversed the trend to align Turkey's visa policy with the EU list, in 2001/2002.
2009	Turkish Citizenship Law, Law No 5901	The law did not change the underlying principles in earlier legislation, but clarified them.
2010	Amendment to the implementing regulation of the law on Work Permits for Foreigners -	– Stricter requirements for employers to employ foreigners – Exemptions were published in April 2011 to ease access for certain categories of foreigner (refugees, victims of trafficking).
2012	Legal change requiring tourists to stay out of Turkey for three months in each six-month period.	LFIP (Law no: 6458) re-states this principle (Article 11). Before, tourists could exit and re-enter the country by renewing their tourist visas. With the legal change, those who want to stay longer than 90 days are required to apply for a short term residence permit.

Date	Policy	Notes on the content / context / intended impact
2012	*Readmission Agreement with the EU*	– The readmission concerns the nationals of the EU Member States and Turkey, plus the third country nationals and the stateless persons who "entered into, or stayed on, the territory of either sides directly arriving from the territory of the other side" (EC, 2013a). – The provision concerning third country nationals and stateless people will come into force in three years. – Turkey signed the RA in exchange for the initiation of EU-Turkey visa liberalisation dialogue.
2013	Adoption of Law on Foreigners and International Protection (Law No. 6458)	– LFIP brings together formerly scattered pieces of legislation on entry, stay and the deportation of foreigners. – For the first time, Turkey's asylum policy is codified as law, as opposed to secondary legislation. – Legal basis for detention and deportation, including procedural guarantees, the right to appeal to decisions on entry bans, detentions and deportations are provided as direct response to ECtHR decisions against Turkey. (Articles 54-55) – Institutionalisation of General Directorate of Migration Administration under the Ministry of Interior (Part V) – Secondary legislation on implementation is under preparation
2013	Circular on the Cost of Health Care of Syrian Asylum Seekers	The circular clarifies that Turkish Prime Ministry Disaster and Emergency Management Authority will undertake cost of health care of Syrians, who should be admitted freely in public hospitals.
2014	Directive on Temporary Protection, Decision no 2014/6883	– The Directive is the first secondary legislation based on LFIP – The Directive concerns nearly two million Syrian refugees fleeing the conflict in Syria since 2011, that are granted Temporary Protection in Turkey.

References

Agartan, T.I. (2012). Marketization and universalism: Crafting the right balance in the Turkish healthcare system. *Current Sociology, 60*(4), 456-471.

Ahmad, A.N. (2008). 'The labour market consequences of human smuggling: 'Illegal' employment in London's migrant economy'. *Journal of Ethnic and Migration Studies, 34*(6), 853-874.

Akalin, A. (2007). 'Hired as a caregiver, demanded as a housewife becoming a migrant domestic worker in Turkey'. *European Journal of Women's Studies, 14*(3), 209-225.

Alami M'chichi, H. (2006). 'La migration dans la coopération UE-Maroc entre tentative de gestion institutionnelle et pragmatisme'. In H.A. M'chichi, B. Hamdouch, & M. Lahlou (eds), *Le Maroc et les Migrations* (pp. 13-40). Rabat Fondation Friedrich Ebert.

Alioua, M. (2008). 'La migration transnationale – logique individuelle dans l'espace national: L'exemple des transmigrants subsahariens à l'épreuve de l'externalisation de la gestion des flux migratoires au Maroc'. *Social Science Information, 47*(4), 697-713.

Alioua, M. (2009). 'Le "passage au politique" des transmigrants subsahariens au Maroc'. In A. Bensaâd (ed.), *Le Maghreb à l'épreuve des migrations subsahariennes. Immigration sur émigration,* (pp. 281-306). Paris: Karthala.

Alioua, M. (2011). *L'étape marocaine des transmigrants subsahariens en route vers l'Europe: l'épreuve de la construction des réseaux et de leurs territoires* (Phd dissertation), Université Toulouse le Mirail-Toulouse II.

Alioua, M. (2013). Régularisation des étrangers au Maroc: Analyse d'une décision historique mais surtout stratégique. Retrieved from http://www.yabiladi.com/articles/details/20944/regularisation-etrangers-maroc-analyse-d-une.html

Amaya-Castro, J. (2011). 'Illegality regimes and the ongoing transformation of contemporary citizenship'. *European Journal of Legal Studies, 4*(2), 137-161.

Ambrosini, M. (2013). *Irregular migration and invisible welfare.* Basingstoke: Palgrave Macmillan.

AMDH (2012a). Communiqué a l'occasion de la journée mondiale des refugiés. Retrieved from http://www.amdh.org.ma/fr/communiques/com-refugies-juin-2012

AMDH (2012b). Situation des migrants Sub-Sahariens au Maroc entre les engagements internationaux et la realité. Rabat: Association Maroccaine d'Etudes et de Recherches sur les Migrations.

AMERM (2008). *De l'afrique subsaharien au Maroc: Les realites de la migration irreguliere.* Rabat: Association Maroccaine d'Etudes et de Recherches sur les Migrations.

Amnesty International (2009). *Stranded refugees in Turkey denied protection.* London: Amnesty International Publications.

Andersson, R. (2014). *Illegality, Inc.: Clandestine migration and the business of bordering Europe.* Oakland, CA: University of California Press.

APDHA (2010). Derechos humanos en la Frontera Sur 2009. Retrieved from Sevilla: http://www.apdha.org/media/informeFS2009.pdf

APDHA (2014). Droits de l'Homme a la Frontiere Sud 2014 Retrieved from Sevilla: http://www.apdha.org/media/frontiere_sud%202014.pdf

Arca, C. (2013). Kayıt dışı istihdam gerçeği ve mücadele yöntemleri. Retrieved from http://www.isvesosyalguvenlik.com/kayit-disi-istihdam-gercegi-ve-mucadele-yontemleri/

Arı, A. (2007). *Türkiye'de yabancı işçiler: Uluslararası göç, işgücü ve nüfus hareketleri* Istanbul: Derin yayınları.

Balta, A. (2010). *The role of NGOs in the asylum system in Turkey: Beyond intermediation.* (MA thesis), Sabancı University, İstanbul.

Basok, T., & Rojas Wiesner, M.L. (2017). 'Precarious legality: Regularizing Central American migrants in Mexico'. *Ethnic and Racial Studies*, 1-20.

Barron, P., Bory, A., Tourette, L., Chauvin, S., & Jounin, N. (2011). *On bosse ici, on reste ici! La grève des sans-papiers: Une aventure inédite.* Paris: La Découverte.

Barron, P., Bory, A., Chauvin, S., Jounin, N., & Tourette, L. (2016). 'State categories and labour protest: Migrant workers and the fight for legal status in France'. *Work, Employment & Society, 30*(4), 631-648.

Belghazi, A. (2015). La GADEM devoile la liste des lieux de detention des migrants au Maroc. Retrieved from http://www.medias24.com/SOCIETE/152908-Le-Gadem-devoile-la-liste-des-lieux-de-detention-des-migrants-au-Maroc.html#sthash.cM7mFIU5.gbpl

Belguendouz, A. (2009). *Le Maroc et la migration irrégulière: Une analyse sociopolitique.* CARIM Analytic and Synthetic Notes; 2009/07; Irregular Migration Series; Socio-political Module.

Berriane, J. (2009). 'Les étudiants subsahariens au Maroc: Des migrants parmi d'autres?' *Méditerranée. Revue géographique des Pays Méditerranéens /Journal of Mediterranean Geography* (113), 147-150.

Berriane, M., & Aderghal, M. (2008). *Etat de la recherche sur les migrations internationales à partir, vers et à travers le Maroc.* Rabat, Morocco: Equipe de Recherche sur la Région et la Régionalisqtion (E3R), Université Mohammed V – Agdal

Berriane, M., Aderghal, M., Amzil, L., & Oussi, A. (2010). *Morocco country and research areas Report.* EUMAGINE Project paper 4.

Betts, A. (2011). *Global migration governance.* Oxford: Oxford University Press.

Biehl, K.S. (2015). 'Spatializing diversities, diversifying spaces: housing experiences and home space perceptions in a migrant hub of Istanbul'. *Ethnic and Racial Studies, 38*(4), 596-607. doi:10.1080/01419870.2015.980293

Biner, Ö. (2014). 'From transit country to host country: A study of transit refugee experience in a border satellite city, Van, Eastern Turkey'. In A.B. Karaçay & A. Üstübici (eds), *Migration to and from Turkey: Changing patterns and shifting policies* (pp. 73-119). İstanbul: ISIS Press.

Bloch, A. (2011). 'Emotion work, shame, and post-Soviet women entrepreneurs: Negotiating ideals of gender and labor in a global economy'. *Identities-Global Studies in Culture and Power, 18*(4), 317-351. doi:10.1080/1070289x.2011.654104

Bloch, A., & McKay, S. (2016). *Living on the margins: Undocumented migrants in a global city.* London: Policy Press.

Bommes, M., & Sciortino, G. (eds) (2011). *Foggy social structures: Irregular migration, European labour markets and the welfare state.* Amsterdam: Amsterdam University Press.

Boswell, C. (2003). 'The 'external dimension' of EU immigration and asylum policy'. *International Affairs, 79*(3), 619-638. doi:10.1111/1468-2346.00326

Brewer, K.T., & Yükseker, D. (2009). 'A survey on African migrants and asylum seekers in Istanbul'. In A. İçduygu & K. Kirişci (eds), *Land of diverse migrations: Challenges of emigration and immigration in Turkey* (pp. 637-724). Istanbul: Bilgi University Press.

Breyer, I., & Dumitru, S. (2007). 'Les sans-papiers et leur droit d'avoir des droits'. *Raisons Politiques* (2), 125-147.

Brian, T., & Laczko, F. (2014). *Fatal journeys. Tracking lives lost during migration.* Geneva: IOM.

Brigden, N., & Mainwaring, Ċ. (2016). 'Matryoshka journeys: Im/mobility during migration'. *Geopolitics, 21*(2), 407-434.

Bröer, C., & Duyvendak, J.W. (2009). 'Discursive opportunities, feeling rules, and the rise of protests against aircraft noise'. *Mobilization: An International Quarterly, 14*(3), 337-356.

Calavita, K. (2005). *Immigrants at the margins: Law, race, and exclusion in Southern Europe.* Cambridge: Cambridge University Press.

Carling, J. (2007). 'Migration control and migrant fatalities at the Spanish-African borders'. *International Migration Review, 41*(2), 316-343. doi:10.1111/j.1747-7379.2007.00070.x

Cassarino, J.-P. (2007). 'Informalising readmission agreements in the EU neighbourhood'. *The International Spectator, 42*(2), 179-196.

Castles, S. (2007). *Comparing the experience of five major emigration countries.* IMI Working Paper Series Paper no:7. Oxford: Oxford University.

Cavatorta, F. (2009). *Civil society activism in Morocco: 'Much ado about nothing'?* Amsterdam: Knowledge Programme Civil Society in West Asia.

Chaudier, J. (2013). Régularisation des migrants au Maroc: Un moratoire sur les expulsions demandé par les associations *yabiladi.com.* Retrieved from http://www.yabiladi.com/articles/details/21232/regularisation-migrants-maroc-moratoire-expulsions.html

Chauvin, S., & Garcés-Mascareñas, B. (2012). 'Beyond informal citizenship: The new moral economy of migrant illegality'. *International Political Sociology, 6*(3), 241-259.

Chauvin, S., & Garcés-Mascareñas, B. (2014). 'Becoming less illegal: Deservingness frames and undocumented migrant incorporation'. *Sociology Compass, 8*(4), 422-432.

Cherti, M., & Grant, P. (2013). *The myth of transit: Sub-Saharan migration in Morocco* London: Institute for Public Policy Research (IPPR).

Chimienti, M. (2011). 'Mobilization of irregular migrants in Europe: A comparative analysis'. *Ethnic and Racial Studies, 34*(8), 1338-1356.

CNDH (2013). Conclusions et recommandations du rapport: "Etrangers et droits de l'Homme au Maroc: Pour une politique d'asile et d'immigration radicalement nouvelle". Retrieved from http://www.ccdh.org.ma/IMG/pdf/Conclusions_et_recommandations_def-2.pdf

Collyer, M. (2007). 'In-between places: Trans-Saharan transit migrants in Morocco and the fragmented journey to Europe'. *Antipode, 39*(4), 668-690.

Collyer, M. (2010). 'Stranded migrants and the fragmented journey'. *Journal of Refugee Studies, 23*(3), 273-293. doi:10.1093/jrs/feq026

Collyer, M., & de Haas, H. (2012). 'Developing dynamic categorisations of transit migration'. *Population, Space and Place, 18*(4), 468-481. doi:10.1002/psp.635

Collyer Michael, Cherti Myriam, Galos Eliza, & Marta, G. (2012). *Responses to irregular migration in Morocco.* London: Institute for Public Policy Research (IPPR).

Coutin, S.B. (1998). 'From refugees to immigrants: The legalization strategies of Salvadoran immigrants and activists'. *International Migration Review,* 901-925.

Coutin, S.B. (2003). *Legalizing moves: Salvadoran immigrants' struggle for US residency.* Ann Arbor, MI: The University of Michigan Press.

Coutin, S.B. (2011). 'The rights of noncitizens in the United States'. *Annual Review of Law and Social Science, 7,* 289-308.

Cvajner, M., & Sciortino, G. (2010). 'Theorizing irregular migration: The control of spatial mobility in differentiated societies'. *European Journal of Social Theory, 13*(3), 389-404. doi:10.1177/1368431010371764

Çelik, A. (2013). 'Trade unions and de-unionization during ten years of AKP rule'. *Perspectives Turkey, 3,* 44-49.

Danış, D., & Parla, A. (2009). 'Nafile soydaşlık: Irak ve Bulgaristan Türkleri örneğinde göçmen, dernek ve devlet'. *Toplum ve Bilim, 114,* 131-158.

Danış, D.A., Taraghi, C., & Pérouse, J.F. (2009). 'Integration in limbo. Iraqi, Afghan, Maghrebi and Iranian migrants in Istanbul'. In A. İçduygu & K. Kirişci (eds), *Land of diverse migrations. Challenges of emigration and immigration in Turkey* (pp. 443-636). Istanbul: İstanbul Bilgi University Press.

Dardağan Kibar, E. (2013). 'An overview and discussion of the new Turkish Law on Foreigners and International Protection'. *Perceptions, 18*(3), 109-128.

De Genova, N. (2002). 'Migrant "illegality" and deportability in everyday life'. *Annual Review of Anthropology, 31*, 419-447.

De Genova, N. (2004). 'The Legal production of Mexican/migrant "illegality"'. *Latino Studies, 2*(2), 160-185.

De Genova, N. (2005). *Working the boundaries: Race, space, and "illegality" in Mexican Chicago.* Durham, NC: Duke University Press.

De Haas, H. (2007). 'Morocco's migration experience: A transitional perspective'. *International Migration, 45*(4), 39-70.

De Haas, H. (2009). 'Country profile: Morocco'. *Focus Migration Country Profile* (16).

De Haas, H. (2014). 'Un siècle de migrations marocaines: Transformations, transitions et perspectives'. In M. Berriane (ed.), *Maroccains de l'Exterieur 2013* (pp. 61-92). Rabat: Fondation Hassan II pour les Marocains Résidant à l'Etranger.

Dedeoğlu, S., & Gökmen, Ç.E. (2010). *Türkiye'de yabancı göçmen kadınların sosyal dışlanması üzerine araştırma* Proje No: 106K258). Muğla Tübitak Proje No: 106K258.

DEMIG (2015) DEMIG POLICY, version 1.3, Online Edition. Oxford: International Migration Institute, University of Oxford. www.migrationdeterminants.eu June 2015

Dimitrovova, B. (2010). 'Re-shaping civil society in Morocco: Boundary setting, integration and consolidation'. *European Integration, 32*(5), 523-539.

Doğan, E., & Genç, S. (2014). 'Impact of visa regimes over travel decisions and patterns of Turkish citizens'. In A. B. Karaçay & A. Üstübici (eds), *Migration to and from Turkey: Changing patterns and shifting policies.* İstanbul: Isis Press.

Dowd, R. (2008). *Trapped in transit: The plight and human rights of stranded migrants.* Genova: UNHCR, Policy Development and Evaluation Service.

Düvell, F. (2013). 'Turkey, the Syrian refugee crisis and the changing dynamics of transit migration'. *IEMed Mediterranean Yearbook 2013*, 278-281.

Düvell, F., & Vollmer, B. (2009). 'Irregular migration in and from the neighbourhood of the EU. A comparison of Morocco, Turkey and Ukraine'. *Clandestino: Undocumented migration. counting the uncountable. Data and trends across Europe.* Centre on Migration, Policy and Society (Compas).

EC (2013a). Cecilia Malmström signs the Readmission Agreement and launches the Visa Liberalisation Dialogue with Turkey. *European Commission Press Release.* Retrieved from http://europa.eu/rapid/press-release_IP-13-1259_en.htm?utm_source=Weekly+Legal+Update&utm_campaign=fdb688b29c-WLU_20_12_2013&utm_medium=email&utm_term=0_7176f0fc3d-fdb688b29c-419648261

EC (2013b). Turkey 2013 Progress Report. Retrieved from http://ec.europa.eu/enlargement/pdf/key_documents/2013/package/brochures/turkey_2013.pdf

EC (2014). *Turkey 2014 Progress Report.* Retrieved from http://ec.europa.eu/neighbourhood-enlargement/sites/near/files/pdf/key_documents/2014/20141008-turkey-progress-report_en.pdf

Elitok, S. (2015). *A step backward for Turkey?: The Readmission Agreement and the hope of visa-Free Europe.* Retrieved from http://ipc.sabanciuniv.edu/wp-content/uploads/2015/12/A-Step-Backward-for-Turkey_The-Readmission-Agreement-and-the-Hope-of-Visa-Free-Europe.pdf

Elmadmad, K. (2007). 'Maroc: La dimension juridique des migrations'. In P. Fargues (ed.), *CARIM Mediterranean Migration Report 2008-2009.* Robert Schuman Centre for Advanced Studies, San Domenico di Fiesole (FI): European University Institute.

Elmadmad, K. (2011). *Rapport sur le cadre juridique et institutionnel de la migration au Maroc Années 2009 et 2010.* CARIM Analytic and Synthetic Notes; 2011/31; Mediterranean and

Sub-Saharan Migration: Recent Developments Series. Retrieved from http://hdl.handle. net/1814/16204

Erder, S. (2007). 'Yabancısız kurgulanan ülkenin "yabancıları"'. In F.A. Arı (ed.), *Türkiye'de yabancı işçiler* (pp. 1-80). İstanbul: Derin Yayınları.

Estévez, A. (2012). *Human rights, migration, and social conflict: Toward a decolonized global justice*. New York: Palgrave Macmillan.

Fargues, P. (2009). 'Work, refuge, transit: An emerging pattern of irregular immigration South and East of the Mediterranean'. *International Migration Review, 43*(3), 544-577.

Fassin, D. (2009). 'Another politics of life is possible'. *Theory, Culture & Society, 26*(5), 44-60. doi:10.1177/0263276409106349

Feliu Martínez, L. (2009). 'Les migrations en transit au Maroc. Attitudes et comportement de la société civile face au phénomène'. *L'Année du Maghreb* (V), 343-362.

Finotelli, C. (2011). 'Regularisation of immigrants in Southern Europe: What can be learned from Spain?' In M. Bommes & G. Sciortino (eds), *Foggy Social Structures: Irregular Migration, European Labour Markets and the Welfare State* (pp. 189-212). Amsterdam: Amsterdam University Press.

FRA (2013). *Fundamental rights at Europe's southern sea borders*. Luxembourg: Publications Office of the European Union.

Frontex (2012). *Annual risk analysis 2012*. Frontex European Agency for the Management of Operational Cooperation at the External Borders of the Member States of the European Union.

Frontex (2014). *Annual risk analysis 2014*. Warsaw: Frontex European Agency for the Management of Operational Cooperation at the External Borders of the Member States of the European Union.

GADEM (2007). *La chasse aux migrants aux frontières Sud de l'UE Conséquence des politiques migratoires européennes L'exemple des refoulements de décembre 2006 au Maroc*. Rabat: GADEM.

GADEM, ALECMA, ARESMA-28, Caminando Fronteras, Chabaka, CCSM, Pateras da Vida (2013). *Rapport sur l'application au Maroc de la Convention internationale sur la protection des droits de tous les travailleurs migrants et des membres de leur famille*. Rabat: GADEM.

GADEM, Allecma, CCSM, & Mission Catholique de Nouadhibou (2014). *Situation des migrants dans le sud du Maroc – Mission d'observation conjointe*. Rabat: GADEM.

Galvin, T.M. (2014). '"We deport them but they keep coming back': The Normalcy of deportation in the daily life of 'undocumented' Zimbabwean migrant workers in Botswana'. *Journal of Ethnic and Migration Studies, 41*(4), 617-634. doi:10.1080/1369183x.2014.957172

Garcés-Mascareñas, B. (2012). *Labour migration in Malaysia and Spain: Markets, citizenship and rights*. Amsterdam: Amsterdam University Press.

Genç, D. (2014). 'Türkiye'nin yabancıların hareketliliğine ilişkin sınır politikası'. In D. Danış & İ. Soysüren (eds), *Sınır ve sınırdışı* (pp. 41-66). İstanbul: Notabene.

Gökmen, Ç.E. (2011). 'Türk turizminin yabancı gelinleri: Marmaris yöresinde turizm sektöründe çalışan göçmen kadınlar'. *Çalışma ve Toplum* (1), 201-231.

Grange, M., & Flynn, M. (2014). *Immigration detention in Turkey*. The Global Detention Project. Lausanne: Global Migration Centre.

Grugel, J., & Piper, N. (2011). 'Global governance, economic migration and the difficulties of social activism'. *International Sociology, 26*(4), 435-454. doi:10.1177/0268580910393043

HCA (2007). *Unwelcome Guests: The detention of refugees in Turkey's "Foreigners' Guesthouses"*. İstanbul: Helsinki Citizens Assembly Refugee Advocacy & Support Program.

Hess, S. (2012). 'De-naturalising transit migration. Theory and methods of an ethnographic regime analysis'. *Population, Space and Place, 18*(4), 428-440.

Honneth, A. (1995). *The Struggle for recognition: The moral grammar of social conflicts* (J. Anderson, Trans.). Cambridge: Polity Press.

HRW (2008). *Stuck in a revolving door Iraqis and other asylum seekers and migrants at the Greece/ Turkey entrance to the European Union.* Human Rights Watch.

HRW (2014). *Abused and expelled Ill-Treatment of Sub-Saharan African migrants in Morocco.* USA: Human Rights Watch.

İçduygu, A. (2006). *Türkiye-Avrupa Birliği ilişkileri bağlamında uluslararası göç tartışmaları.* İstanbul: TÜSİAD Yayınları.

İçduygu, A. (2007). 'EU-ization matters: Changes in immigration and asylum practices in Turkey'. In T. Faist & A. Ette (eds), *The Europeanization of National Policies and Polities of Immigration* (pp. 201-222). London: Palgrave MacMillan Publishers.

İçduygu, A. (2011). *The Irregular migration corridor between the EU and Turkey: Is it possible to block it with a Readmission Agreement?* EU-US Immigration Systems 2011/14 Research Report Case Study Robert Schuman Centre for Advanced Studies, San Domenico di Fieso le (FI): European University Institute.

İçduygu, A., & Aksel, D.B. (2012). *Irregular migration in Turkey.* Ankara: IOM.

İçduygu, A., & Kirişci, K. (eds) (2009). *Land of diverse migrations: Challenges of emigration and immigration in Turkey.* İstanbul: Bilgi University Press.

İçduygu, A., & Sert, D. (2009). Country profile Turkey. *Focus Migration Country Profile no: 5.*

İçduygu, A., Soyarık, N., Korfalı, D.K., Çoban, C., & Gökçe, A. (2014). *Türkiye'de göçmenlerin vatandaş olma eğilimleri: Algılar ve deneyimler.* İstanbul: Tübitak PROJE NO: 112K168.

İçduygu, A., & Üstübici, A. (2014). 'Negotiating mobility, debating borders: Migration diplomacy in Turkey-EU Relations'. In H. Schwenken & S. Ruß-Sattar (eds), *New border and citizenship Politics* (pp. 44-59). London: Palgrave Macmillan.

İçduygu, A., & Yükseker, D. (2012). 'Rethinking transit migration in Turkey: reality and re-presentation in the creation of a migratory phenomenon'. *Population, Space and Place, 18*(4), 441-456.

IHAD (2009). *İltica ve sığınma hakkı 2008 izleme raporu* Ankara: Human Rights Research Asssociation.

İHAD (2012). *2011 İltica ve sığınma hakkı 2011 izleme raporu.* Ankara: Human Rights Research Asssociation.

Ihlamur-Öner, S.G. (2013). 'Turkey's refugee regime stretched to the limit? The case of Iraqi and Syrian refugee flows'. *Perceptions, 18*(3), 191-228.

Iskander, N. (2010). *Creative state: Forty years of migration and development policy in Morocco and Mexico.* Ithaca, NY: Cornell University Press.

Jacobs, A. (2012). *Sub-Saharan irregular migration in Morocco; the politics of civil society and the state in the struggle for migrants' rights.*: Fulbright Morocco, September 2011-August 2012.

Kalir, B. (2012). 'Illegality rules: Chinese migrant workers caught up in the illegal but licit operations of labour migration regimes'. In B. Kalir & M. Sur (eds), *Transnational Flows and Permissive Polities* (pp. 27-54). Amsterdam: Amsterdam University Press.

Karakayali, S., & Rigo, E. (2010). 'Mapping the European space of circulation'. In N. De Genova & N. Peutz (eds), *The deportation regime: Sovereignty, space, and the freedom of movement* (pp. 123-144). Durham, NC: Duke University Press.

Kastner, K. (2010). 'Moving relationships: Family ties of Nigerian migrants on their way to Europe'. *African and Black Diaspora: An International Journal, 3*(1), 17-34.

Kaşka, S. (2009). 'The new international migration and migrant women in Turkey: The case of Moldovan domestic workers'. In A.İ.a.K. Kirişci (ed.), *Land of diverse migrations: Challenges of emigration and immigration in Turkey* (pp. 725-788). İstanbul: İstanbul Bilgi University Pres.

Kaşlı, Z. (2016). 'Who do migrant associations represent? The role of 'ethnic deservingness' and legal capital in migrants' rights claims in Turkey'. *Journal of Ethnic and Migration Studies, 42*(12), 1996-2012. doi:10.1080/1369183X.2015.1137753

Kavak, S. (2016). 'Syrian refugees in seasonal agricultural work: A case of adverse incorporation in Turkey'. *New Perspectives on Turkey, 54,* 33-53.

Keough, L.J. (2006). 'Globalizing 'postsocialism:' Mobile mothers and neoliberalism on the margins of Europe'. *Anthropological Quarterly, 79*(3), 431-461.

Khachani, M. (2011). *La question migratoire au Maroc : Données récentes.* CARIM Analytic and Synthetic Notes; 2011/71; Mediterranean and Sub-Saharan Migration: Recent Developments Series; Socio-political Module.

Kılıç, T. (2014). 'Batı sınırından doğu sınırına: Geri Kabul Anlaşması, "push back" ve Özbek mülteciler'. In D. Danış & İ. Soysüren (eds), *Sınır ve sınırdışı: Türkiye'de yabancılar, göç ve devlete disiplinlerarası bakışlar* (pp. 427-449). İstanbul: Notabene.

Kimball, A. (2007). *The Transit state: A comparative analysis of Mexican and Moroccan immigration policies.* San Diego, CA: University of California.

Kirişci, K. (2008). 'Migration and Turkey: The dynamics of state, society and politics'. In R. Kasaba (ed.), *The Cambridge history of Turkey: Turkey in the modern world* (pp. 175-198). New York: Cambridge University Press.

Kirişci, K. (2009). *Mirage or reality: Post-national Turkey and its implication for immigration.* EUI RSCAS; 2009/14; CARIM Research Report.

Kirişci, K. (2012). 'Turkey's new draft law on asylum: What to make of it?' In S.P. Elitok & T. Straubhaar (eds), *Turkey, migration and the EU: Potentials, challenges and opportunities* (pp. 63-84). Hamburg: Hamburg University Press.

Kohli, A., Evans, P., Katzenstein, P.J., Przeworski, A., Rudolph, S.H., Scott, J.C., & Skocpol, T. (1995). 'The role of theory in comparative politics: A symposium'. *World Politics, 48*(01), 1-49.

Korkmaz, E.E. How do Syrian refugee workers challenge supply chain management in the Turkish garment industry? *University of Oxford IMI Working Paper, 133.*

Kubal, A. (2013). 'Conceptualizing semi-legality in migration research'. *Law & Society Review, 47*(3), 555-587. doi:10.1111/lasr.12031

Landman, T. (2003). *Issues and methods in comparative politics: An introduction.* London and New York: Routledge.

Laubenthal, B. (2007). 'The emergence of pro-regularization movements in Western Europe'. *International Migration, 45*(3), 101-133.

Lemaizi, S. (2013). 'Regularisation des migrants: Le plus dur reste à faire'. *L'observateur du Maroc et de l'Afrique.* Retrieved from http://lobservateurdumaroc.info/2013/12/04/regularisation-migrants-dur-reste-faire/

Lemke, T. (2007). 'An indigestible meal? Foucault, governmentality and state theory'. *Distinktion: Scandinavian Journal of Social Theory, 8*(2), 43-64.

Lentin, R., & Moreo, E. (2015). 'Migrant deportability: Israel and Ireland as case studies'. *Ethnic and Racial Studies, 38*(6), 894-910. doi:10.1080/01419870.2014.948477

Lipsky, M. (1980). *Street-Level bureaucracy: The dilemmas of individuals in public service.* New York: Russell Sage Foundation.

Lutterbeck, D. (2006). 'Policing migration in the Mediterranean: ESSAY'. *Mediterranean Politics, 11*(1), 59-82.

Marrow, H.B. (2009). 'Immigrant bureaucratic incorporation: The dual roles of professional missions and government policies'. *American Sociological Review, 74*(5), 756-776.

Mazlumder (2014). *Kamp dışında yaşayan Suriyeli kadın sığınmacılar raporu.* İstanbul: Mazlumder.

McNevin, A. (2006). 'Political belonging in a neoliberal era: The struggle of the Sans-Papiers'. *Citizenship Studies, 10*(2), 135-151.

McNevin, A. (2012). 'Undocumented citizens? Shifting grounds of citizenship in Los Angeles'. In P. Nyers & K. Rygiel (eds), *Citizenship, migrant activism and the politics of movement* (pp. 165-183). New York and London: Routledge.

Menjívar, C. (2006). 'Liminal legality: Salvadoran and Guatemalan immigrants' lives in the United States'. *American Journal of Sociology, 111*(4), 999-1037.

Menjívar, C. (2014). 'Immigration law beyond borders: Externalizing and internalizing border controls in an era of securitization'. *Annual Review of Law and Social Science, 10*, 353-369.

Menjívar, C., & Coutin, B. (2014). 'Challenges of recognition, participation, and representation for the legally liminal: A comment'. In T.-D. Truong, D. Gasper, J. Handmaker, & S.I. Bergh (eds), *Migration, gender and social justice-perspectives on human insecurity. Hexagon Series on Human and Environmental Security and Peace* (pp. 351-364). Heidelberg: Springer.

Menjívar, C., & Kanstroom, D. (2014). *Constructing illegality in America: Immigrant experiences, critiques, and resistance.* New York: Cambridge University Press.

Mghari, M. (2009). 'Maroc: La dimension démographique et économique des migrations'. In P.E. Fargues (ed.), *CARIM Mediterranean Migration Report 2008-2009.* Robert Schuman Centre for Advanced Studies, San Domenico di Fiesole (FI): European University Institute.

Migreurop (2006). *Guerre aux migrants – Le Livre Noir de Ceuta et Melilla.* Paris: Migreurop.

Ministry in Charge of Moroccans Abroad and Migration Affairs (2016). *Politique Nationale d'Immigration et d'Asile: 2013-2016.* Rabat: Ministry in Charge of Moroccans Abroad and Migration Affairs.

Moore, S.F. (1973). 'Law and social change: The semi-autonomous social field as an appropriate subject of study'. *Law and Society Review*, 719-746.

Moroccan Ministry of Health (2014a). *Axe I: Spécificités d'intégration dans le domaine de la santé* Paper presented at the Séminaire international sur l'intégration des immigrés au Maroc Nouvelle politique Migratoire quelle stratégie d'intégration, 10-11 March 2014, Rabat, Morocco.

Moroccan Ministry of Health (2014b). *Plan d'action intégré pour la promotion de la santé des populations migrantes en situation administrative irrégulière au Maroc.* Paper presented at the Assises Nationales de l'ALCS, 17-18-19 January 2014, Rabat, Maroc.

Mountz, A., & Loyd, J.M. (2014). 'Constructing the Mediterranean region: obscuring violence in the bordering of Europe's migration "crises"'. *ACME: An International E-Journal for Critical Geographies, 13*(2), 173-195.

MSF (2010). *Sexual violence and migration. The hidden reality of Sub-Saharan women trapped in Morocco en route to Europe.* Rabat: Médecins Sans Frontières.

MSF (2013). *Violence, vulnerability and migration: Trapped at the gates of Europe.* Rabat: Médecins Sans Frontières.

Natter, K. (2014). 'The formation of Morocco's policy towards irregular migration (2000-2007): Political rationale and policy processes'. *International Migration, 52*(5), 15-28. doi:10.1111/imig.12114

Nicholls, W. (2013). *The dreamers: How the undocumented youth movement transformed the immigrant rights debate.* Stanford, CA: Stanford University Press.

Nicholls, W.J. (2014). 'From political opportunities to niche-openings: The dilemmas of mobilizing for immigrant rights in inhospitable environments'. *Theory and Society, 43*(1), 23-49.

Nieselt, T. (2014). *Becoming the "Gendarme of Europe" in return for more mobility? An analysis of the added value for Morocco to conclude a Mobility Partnership with the European Union* (Bachelor in Public Administration (BA) Bachelor in European Studies (BSc)), Westfälische Wilhelms-Universität Münster, Institute of Political Science University of Twente, School of Management and Governance.

Nyers, P., & Rygiel, K. (2012). 'Introduction. Citizenship, migrant activism and the politics of movement'. In P. Nyers & K. Rygiel (eds), *Citizenship, migrant activism and the politics of movement* (pp. 1-20). New York: Routledge.

Oelgemöller, C. (2011). "'Transit' and 'suspension': Migration management or the metamorphosis of asylum-seekers into 'illegal' immigrants'. *Journal of Ethnic and Migration Studies, 37*(3), 407-424.

Ozcurumez, S., & Şenses, N. (2011). 'Europeanization and Turkey: studying irregular migration policy'. *Journal of Balkan and Near Eastern Studies, 13*(2), 233-248.

Ozcurumez, S., & Yetkin, D. (2014). 'Limits to regulating irregular migration in Turkey: What constrains public policy and why?' *Turkish Studies, 15*(3), 442-457.

Ozmenek, E. (2001). 'UNHCR in Turkey'. *Refuge: Canada's Journal on Refugees, 19*(5), 54-61.

Özden, S. (2013). *Syrian refugees in Turkey*. MPC Research Reports 2013/05. Robert Schuman Centre for Advanced Studies, San Domenico di Fiesole (FI): European University Institute.

Özgür, N., & Özer, Y. (2010). *Türkiye'de sığınma sisteminin Avrupalılaştırılması*. İstanbul: Derin Yayınları.

Pallister-Wilkins, P. (2015). 'The humanitarian politics of European border policing: Frontex and border police in Evros'. *International Political Sociology, 9*(1), 53-69.

Papadopoulos, D., Stephenson, N., & Tsianos, V. (2008). *Escape routes: Control and subversion in the 21st century*. Ann Arbor, MI: Pluto Press.

Parla, A. (2011). 'Labor migration, ethnic kinship, and the conundrum of citizenship in Turkey'. *Citizenship studies, 15*(3-4), 457-470.

Però, D., & Solomos, J. (2010). 'Introduction: Migrant politics and mobilization: Exclusion, engagements, incorporation'. *Ethnic and Racial Studies, 33*(1), 1-18.

Perrin, D. (2011). *Country Report: Morocco*. EUDO Citizenship Obvervatory. RSCAS, European University Institute.

Peutz, N., & De Genova, N. (2010). 'Introduction'. In N. De Genova & N. Peutz (eds), *The deportation regime: Sovereignty, space, and the freedom of movement* (pp. 1-32). Durham, NC: Duke University Press.

Pian, A. (2009). *Aux nouvelles frontières de l'Europe: L'aventure incertaine des Sénégalais au Maroc*. Paris: La Dispute.

Pickerill, E. (2011). 'Informal and entrepreneurial strategies among sub-Saharan migrants in Morocco'. *The Journal of North African Studies, 16*(3), 395-413.

Qassemy, H. (2014). *Les enfants migrants et l'école marocaine Etat des lieux sur l'accès à l'éducation des enfants migrants subsahariens au Maroc*. Rabat: Tamkine Migrants.

Raissiguier, C. (2014). 'Troubling borders: *Sans papiers* in France'. In H. Schwenken & S. Ruß-Sattar (eds), *New border and citizenship politics* (pp. 156-170). London: Palgrave Macmillan.

Sadiq, K. (2008). *Paper citizens: How illegal immigrants acquire citizenship in developing countries*. New York: Oxford University Press.

Samers, M. (2004). 'An emerging geopolitics of 'illegal' immigration in the European Union'. *European Journal of Migration and Law, 6*(1), 27-45.

Sartori, G. (1991). 'Comparing and miscomparing'. *Journal of Theoretical Politics, 3*(3), 243-257.

Sater, J.N. (2007). *Civil society and political change in Morocco*. New York: Routledge.

Scheel, S., & Ratfisch, P. (2014). 'Refugee protection meets migration management: UNHCR as a global police of populations'. *Journal of Ethnic and Migration Studies, 40*(6), 924-941.

Semeraro, G. (2011). *Migration effects on civil society and institutional landscape: the case of Morocco* (MA thesis), Utrecht University, Utrecht.

Shore, C., & Wright, S. (2003). 'Policy: A new field of anthropology'. In C. Shore & S. Wright (eds), *Anthropology of policy: Perspectives on governance and power* (pp. 3-33). London: Routledge.

SRHRM (2013). *Report by the Special Rapporteur on the human rights of migrants, François Crépeau, Mission to Turkey (25-29 June 2012)*, Human Rights Council. A/HRC/23/46/Add.2. 17 April 2013.

Suter, B. (2012). *Tales to transit: Sub-Saharan African migrants' experiences in Istanbul*. (PhD dissertation), Linköping University Malmö University, Malmö.

Şenses, N. (2012). *Rights and democratic accountability: A comparative study on irregular migration in Greece, Spain, Turkey* (PhD dissertation), Bilkent University, Ankara.

Şimşek, D. (2015). 'Suriyeli sığınmacılar ve "misafir" olma hali'. *Birikim, 311* (Mart 2015), 55-62.

Taran, P.A., & Geronimi, E. (2003). *Globalization, labour and migration: Protection is paramount*: International Labour Office.

Tes-İş (October 2005). *Çalışma Hayatında Göçmenler*. Ankara: Tes-İş.

Toksöz, G., Erdoğdu, S., & Kaşka, S. (2012). *Irregular migration in Turkey and the situation of migrant workers in the labour market*. Ankara: IOM.

Toksöz, G., & Ünlütürk Ulutaş, Ç. (2012). 'Is Migration feminized? A gender- and ethnicity-based review of the literature on irregular migration to Turkey'. In S.P. Elitok & T. Straubhaar (eds), *Turkey, migration and the EU: Potentials, challenges and opportunities* (pp. 85-112). Hamburg: Hamburg University Press.

Tolay, J. (2012). 'Turkey's "Critical Europeanization": Evidence from Turkey's immigration policies'. In S.P. Elitok & T. Straubhaar (eds), *Turkey, migration and the EU: Potentials, challenges and opportunities* (pp. 39-62). Hamburg: Hamburg University Press.

Trauner, F., & Deimel, S. (2013). 'The impact of EU migration policies on African countries: The case of Mali'. *International Migration, 51*(4), 20-32.

Tsianos, V., & Karakayali, S. (2010). 'Transnational migration and the emergence of the European border regime: An ethnographic analysis'. *European Journal of Social Theory, 13*(3), 373-387. doi:10.1177/1368431010371761

Turkish Parliament (2010). Türkiye'de bulunan mülteciler, sığınmacılar, ve yasa dışı göçmenlerin sorunlarını inceleme raporu. Ankara: Turkish Parliament Human Rights Inquiry Committee.

Turkish Parliament (2012). Ülkemize sığınan Suriye vatandaşlarının barındıkları çadırkentler hakkında inceleme raporu-2-. Ankara: Turkish Parliament Human Rights Inquiry Committee.

Turkish Parliament (2014). Edirne ili yasa dışı göç inceleme raporu. Ankara: Turkish Parliament Human Rights Inquiry Committee.

Tyler, I., & Marciniak, K. (2013). 'Immigrant protest: An introduction'. *Citizenship Studies, 17*(2), 143-156.

Ulusoy, O., & Kılınç, U. (2014). 'Yabancıların sınırdışı işlemlerinde AİHM İçtüzük Kural 39 ve Anayasa Mahkemesi bireysel başvuru yolları'. In D. Danış & İ. Soysüren (eds), *Sınır ve sınırdışı* (pp. 246-272). İstanbul: Notabene Yayınları.

UNHCR (2014). UNHCR concerned over attempt to legalize automatic returns from Spanish enclaves. *Briefing Notes, 28 October 2014*. Retrieved from http://www.unhcr.org/544f71e96. html

Üstübici, A. (2015). 'Dynamics in emigration and immigration policies of Morocco: A double engagement'. *Migration and Development, 4*(2), 238-255.

Valluy, J. (2007). 'Le HCR au Maroc: Acteur de la politique européenne d'externalisation de l'asile'. *L'Année du Maghreb [en ligne], 3*. doi:10.4000/anneemaghreb.398

Van der Leun, J. (2003). *Looking for loopholes: Processes of incorporation of illegal immigrants in the Netherlands*. Amsterdam: Amsterdam University Press.

Van Meeteren, M. (2012). 'Transnational activities and aspirations of irregular migrants in Belgium and the Netherlands'. *Global Networks, 12*(3), 314-332. doi:10.1111/j.1471-0374.2012.00354.x

Villegas, P.E. (2014). '"I can't even buy a bed because I don't know if I'll have to leave tomorrow": Temporal orientations among Mexican precarious status migrants in Toronto'. *Citizenship Studies, 18*(3-4), 277-291.

Willen, S.S. (2007a). 'Toward a critical phenomenology of "illegality": State power, criminalization, and abjectivity among undocumented migrant workers in Tel Aviv, Israel'. *International Migration, 45*(3), 8-38. doi:10.1111/j.1468-2435.2007.00409.x

Willen, S.S. (2007b). 'Exploring "illegal" and "irregular" migrants' lived experiences of law and state power'. *International Migration, 45*(3), 2-7.

Wilmes, M. (2011). 'Irregular migration and foggy organisational structures: Implications of a German city study'. In M. Bommes & G. Sciortino (eds), *Foggy social structures irregular migration, European labour markets and the welfare state* (pp. 117-140). Amsterdam: Amsterdam University Press.

Wissink, M., Düvell, F., & Van Eerdewijk, A. (2013). 'Dynamic migration intentions and the impact of socio-institutional environments: A transit migration hub in Turkey'. *Journal of Ethnic and Migration Studies, 39*(7), 1087-1105.

Wolff, S. (2008). 'Border management in the Mediterranean: Internal, external and ethical challenges'. *Cambridge Review of International Affairs, 21*(2), 253-271.

Wunderlich, D. (2010). 'Differentiation and policy convergence against long odds: Lessons from implementing EU migration policy in Morocco'. *Mediterranean Politics, 15*(2), 249-272.

Wunderlich, D. (2013). 'Implementing EU external migration policy: Security-driven by default&quest'. *Comparative European Politics, 11*(4), 406-427.

Yılmaz, A. (2014). 'Türkiye'de yabancıların sınırdışı edilmesi: Uygulama ve yargısal denetim'. In D. Danış & İ. Soysüren (eds), *Sınır ve sınırdışı* (pp. 207-245). İstanbul: Notabene Yayınları.

Yılmaz, G. (2012). *The Impact of foreign nationals on state policy: Refugees and asylum seekers, European Court of Human Rights case law and Turkish asylum law* (MSc thesis), Boğaziçi University, İstanbul.

Yükseker, D. (2004). 'Trust and gender in a transnational market: the public culture of Laleli, Istanbul'. *Public Culture, 16*(1), 47-66.

Yükseker, D., & Brewer, K.T. (2011). 'Astray and stranded at the gates of the European Union: African transit migrants in Istanbul'. *New Perspectives on Turkey, 44*, 129-160.

Zapata-Barrero, R., & Witte, N.D. (2007). 'The Spanish governance of EU borders: Normative questions'. *Mediterranean Politics, 12*(1), 85-90.

Index